Project Based Problem Solving and Decision Making

Project Based Problem Solving and Decision Making

A Guide for Project Managers

Harold Kerzner
Senior Executive Director for Project Management
International Institute for Learning, Inc. (IIL), USA

Published by John Wiley & Sons, Inc., Hoboken, New Jersey.
Published simultaneously in Canada.

For general information on our other products and services or for technical support, please contact our Customer Care Department within the United States at (800) 762-2974, outside the United States at (317) 572-3993 or fax (317) 572-4002.

Wiley also publishes its books in a variety of electronic formats. Some content that appears in print may not be available in electronic formats. For more information about Wiley products, visit our web site at www.wiley.com.

Library of Congress Cataloging-in-Publication Data Applied for:

Hardback ISBN: 9781394207831

Cover Design: Wiley
Cover Image: © Konstantin Faraktinov/Shutterstock

Set in 9.5/12.5 STIX Two Text by Straive, India

SKY10056799_100423

Contents

Preface

The environment in which the project managers perform has changed significantly in the past three years due to COVID-19 pandemic and other factors. Our projects have become more complex. There are new internal and external forces that now impact how problems are solved. The importance of time and cost has reached new heights in the minds of clients and stakeholders. Clients want to see the value in the projects they are funding. All of this is creating challenges for project managers in how they identify and resolve problems. To make matters more complex, project managers are now seen as managing part of a business when managing a project and are expected to make both project and business decisions.

Decisions are no longer a single-person endeavor. Project managers are expected to form problem-solving and decision-making teams. Most project managers have never been trained in problem-solving, brainstorming, creative thinking techniques, and decision-making. They rely on experience as the primary teacher. While that sounds like a reasonable approach, it can be devastating if project managers end up learning from their own mistakes rather than the mistakes of others. It is a shame that companies are unwilling to invest even small portions of their training budgets in these courses.

There are numerous books available on problem-solving and decision-making. Unfortunately, they look at the issues from a psychological perspective with applications not always relevant to project and program managers. What I have attempted to do with this book is extract the core concepts of problem-solving and decision-making that would be pertinent to project managers and assist them with their jobs.

Some books use the term problem analysis rather than problem-solving. Problem analysis can be interpreted as simply looking at the problem and gathering the facts, but not necessarily developing alternative solutions for later decision-making. In this book, problem-solving is used throughout reflecting the identification of alternatives as well.

Hopefully, after reading this book, you will have a better understanding and appreciation for problem-solving and decision-making.

Seminars and webinars on project management, problem-solving, and decision-making can be arranged by contacting

Lori Milhaven, CSPO
Executive Vice President, Strategic Programs
International Institute for Learning

Harold Kerzner
Senior Executive Director for Project Management,
International Institute for Learning, Inc. (IIL), USA
September 2023

About the Companion Website

This book is accompanied by a companion website:

www.wiley.com/go/kerzner/projectbasedproblemsolving

The website includes instructor manual and chapter wise PowerPoints with extracts from the book for instructors

1

Understanding the Concepts

1.0 The Necessity for Problem-Solving and Decision-Making

We are forced to make decisions in our everyday lives. We must decide what to eat, how to dress, where to go, when to go, and even who to socialize with. We may make thirty or more decisions a day. Some decisions, such as personal investment decisions, may be critical, whereas other decisions may be just routine. Most of these types of decisions we make by ourselves and usually have confidence in the fact that we made the right decision. And for some of the decisions, we can expend a great deal of time thinking through them.

But once we get to our place of employment, the decision-making process changes. We often must involve many people in the process; some of whom we may never have met or worked with previously. The outcome of the decision can affect a multitude of people, many of whom may be unhappy with the outcome. The risks of a poor decision can lead to catastrophic consequences for the business. People that are unhappy with the decision and do not understand it may view you now as an enemy rather than as a friend.

When we make personal decisions, we usually adopt a "let's live with it" attitude. If the decision is wrong, we may try to change it. But in a business environment, there may be a significant cost associated with changing a decision. Some business decisions are irreversible.

But there is one thing, we know for sure in a business environment: anybody that always makes the right decision probably is not making enough decisions. Expecting to always make the right decision is wishful thinking.

Problem-solving and decision-making go hand-in-hand. Decisions are made when we have issues accompanied by choices to make. In general, we must have a problem prior to making decisions. But there is a strong argument that an

Project Based Problem Solving and Decision Making: A Guide for Project Managers,
First Edition. Harold Kerzner.

Companion Website: www.wiley.com/go/kerzner/projectbasedproblemsolving

understanding of decision-making is needed and used as part of identifying the problem and developing alternatives. This is why most books discuss problem-solving and decision-making together.

1.1 Problems and Opportunities[1]

Problem-solving usually begins with the identification of a problem. A problem is a question raised for inquiry, consideration, or in need of a solution. Failure to meet product quality standards during manufacturing is a problem. Inventory shortages during a production run are a problem. Having resources assigned to a project that lack the necessary skills is a problem. These problems, if viewed positively, can become an opportunity for the astute manager rather than just a problem.

Not all problems require solutions. If you lack sufficient resources to maintain a schedule, senior management may allow the schedule to slip rather than hiring additional resources or reassigning resources from other projects that may have a higher priority.

An opportunity is a favorable chance for advancement or progress. If a company's manufacturing process fails to produce a quality product, then the opportunity exists to improve product quality through

- Review of the product design process
- Analysis of manufacturing engineering standards
- Assessment of quality inspection techniques
- Evaluation of adequacy of manufacturing management
- Investigation of motivation and commitment of manufacturing personnel to implement adequate quality policies and procedures.

The difference between a problem and an opportunity depends on the beholder. However, problems and opportunities should be differentiated. David B. Gleicher, a management consultant, distinguished between the two terms in the following way: A problem is "something that endangers the organization's ability to reach its objectives, while an opportunity is something that offers the chance to exceed objectives."[2]

Peter Drucker made it clear that opportunities rather than problems are the keys to organizational and managerial success. He observed that solving a problem merely restores normality; but results must come from the exploitation of opportunities. He linked exploitation of opportunities to finding the right things to do and concentrating "resources and efforts on them."[3]

1 This section has been adapted from Cleland and Kerzner (1986).

2 Cited in Stoner (1982).

3 Drucker (1964).

Identifying problems or opportunities is a key activity of all managers, including project managers. Successful managers do not wait for someone else to tell them what to do; they must find ways to figure that out for themselves.

There are early warning signs and situations that can alert managers to possible problems (issues or opportunities). First, when the project team is expected to perform differently than they did in the past; second, when problems occur resulting from a deviation from an existing plan and no previous history exists; and third, when competitors are outperforming your organization.

1.2 Research Techniques in the Basic Decision-Making Process

Human performance in the way we address problems and make decisions has been the subject of active research from several perspectives. There has been significant research in this area, and most results are part of four perspectives. The four basic perspectives are:

- The psychological perspective
- The cognitive perspective
- The normative perspective
- The problem-solving perspective.

From a psychological perspective, it is necessary to examine individual decisions in the context of a set of needs, preferences an individual has, and values desired. From a cognitive perspective, the decision-making process must be regarded as a continuous process integrated in the interaction with the environment. From a normative perspective, the analysis of individual decisions is concerned with the logic of decision-making and rationality as well as the invariant choices it leads to.

Yet, at another level that is perhaps more closely aligned with a project environment, it might be regarded as a problem-solving activity which is terminated when a satisfactory solution is found. Therefore, decision-making is a reasoning and/or emotional process which can be rational or irrational and can be based on explicit assumptions or tacit assumptions. It is often impossible to separate discussions of decision-making from problem-solving. Yet both involve selecting among alternatives. The focus of this book will be the problem-solving perspective, although, in a project management environment, we could argue that all four perspectives somehow interact in the way we make decisions.

1.3 Facts About Problem-Solving and Decision-Making

There are several facts or generalities that we consider when discussing problem-solving and decision-making:

- Businesses today are much more complex than before and so are the problems and decisions that must be made.
- Problem-solving techniques are used not only to solve problems but also to take advantage of opportunities.
- Today, we seem to be flooded with information to the point where we have information overload and cannot discern what information is actually needed or useful for solving problems.
- Lower to middle levels of management are often valuable resources to have when discussing the technical side of problems. Senior management is valuable in the knowledge of how the problem (and its solution) relates to the overall business and the impact of the enterprise environmental factors.
- Problem-solving today is a core competency, yet most companies provide very little training for their employees on problem-solving and decision-making.
- The project team may be composed of numerous subject matter experts, but the same people may not be creative and capable of thinking creatively about solving problems.
- The people who created the problem may not be capable of solving the it.
- Few people seem to know the relationship between creativity and problem-solving.

1.4 Who Makes the Decision?

Problem-solving and decision-making may not be performed by the same person. As an example, the project manager may ask the project team to assist in identifying alternatives for resolving a problem, or the project manager may perform the activities alone. However, the final decision on which alternatives should be taken may be made by executives, the project team, functional managers, or stakeholders. As such, when discussing decision-making in this book, we use the word "manager" as the person making the decision, and it could represent different individuals other than the project manager. Usually, the people involved in the decision are the ones who will be affected by the outcome.

1.5 Information Overload

Today, there seems to be an abundance of information available to everyone. We all seem to suffer from information overload thanks to advances in information system technologies. Our main problem is being able to discern what information is critical and what information should be discarded or stored in archives.

For simplicity's sake, information can be broken down into primary and secondary information. Primary information is information that is readily available to us. This is information that we can directly access from our desktop or laptop. Information that is company sensitive or considered proprietary information may be password protected but still accessible.

Secondary information is information that must be collected from someone else. Even with information overload, project managers generally do not have all the information they need to solve a problem and make a timely decision. This is largely due to the complexity of our projects as well as the complexity of the problems that need to be resolved. We generally rely on a problem-solving team to provide us with the secondary information. The secondary information is often more critical for decision-making than the primary information. Many times, the secondary information is controlled by the subject matter experts, and we must rely upon them to tell us what information is directly pertinent to this problem.

Collecting the information, whether primary or secondary, can be time-consuming. Information overload often forces us to spend a great deal of time searching through information when this time should be spent on problem-solving.

1.6 Getting Access to the Right Information

The project manager's challenge is not just getting the information but getting the right information in a timely manner. Sometimes, the information that the project manager needs, especially secondary information, is retained by people that are not part of the project or the problem-solving team. An example might be information related to politics, stakeholder relations management, economic conditions, cost of capital, and other enterprise environmental factors. This information may be retained by senior management or stakeholders.

Because timing is essential, project managers should have the right to talk directly with anyone they need to converse with to obtain the necessary information to solve a problem. Having to always go through the chain of command to access the information creates problems and wastes valuable time. Information is often seen as a source of power, which is one of the reasons why sometimes the chain of command must be followed in some companies.

Behind every door in a company is information of some sort. Project managers must be able to open those doors as needed. If project managers do not have access to those doors, then there are two options: follow the chain of command and hope that the information is not filtered by the time it gets to you, or invite the person with this information to attend the problem-solving meeting. Based on where the person with the information resides in the organizational hierarchy, their availability or willingness to attend the meeting is determined. The higher up they reside, the less likely they will be able to attend your meeting in the near term. The project manager's accessibility to information is critical.

1.7 The Lack of Information

Even though we have information overload and access to secondary information sources, there is no guarantee we will have readily available all of the information we need. People that need to make decisions must accept the fact that they generally will not have all of the information they need on hand. This can happen at all levels of management, not just on projects. We must be willing to make the best possible decisions based upon the information we have at that time, even if it is partial information.

Too often, we rely on the chain of command for getting the information to help resolve a problem. If people believe that "possessing information is power," access to the needed information can be a problem especially if they withhold some of the information. Because of the criticality of the project's constraints, time is not necessarily a luxury. Project managers must have the right or authority to access those who possess the information. This assumes, of course, the project manager knows where the information resides. This is sometimes the greater challenge, especially if the needed information is nowhere to be found within the company. We must go outside the company to get the critical information.

Problem-solving is most frequently based upon the best available information. Having all the information needed to make a decision is wishful thinking.

Discussion Questions

1. Why must problem-solving and decision-making be discussed together?
2. How do you differentiate between a problem and an opportunity? Can guidelines be established for differentiation?
3. When competitors are outperforming your organization, is this a problem, an opportunity, or possibly both?

4. Problem solving is considered as a "core competency" for future project managers. Why has this recognition not happened sooner?
5. In the future, will project managers be expected to make more or fewer decisions by themselves, and why?
6. What are the differences between primary and secondary information?
7. Why have project managers been challenged by not being able to get access to the right information in a timely manner?

References

Cleland, D.I. and Kerzner, H. (1986). *Engineering Team Management.* New York: Van Nostrand Reinhold, p. 229.

Drucker, P.F. (1964). *Managing for Results.* New York: Harper & Row, p. 5.

Stoner, J.A.F. (1982). *Management*, 2. Englewood Cliffs, N.J.: Prentice-Hall, pp. 166–167.

2

Understanding the Project Environment and the Impact on Problem Solving

2.0 Understanding the Project Environment[1]

There are several activities that project teams must perform. Two key activities carried out in the management of a team include: deciding what needs to be done and then doing it. In this chapter we deal with the decision-making process when managing within a project environment.

A decision is the act or process of selecting a course of action after consideration of the alternative ways that resources can be used to attain organization objectives. In the process of making a decision on traditional projects, the decision-maker—usually a team member or a subject matter expert assigned to the team—carries out certain activities:

- Evaluation of the current environment and situation
- Assessment of what the situation and environment will be when the decision will be implemented
- Analysis of how the decision will affect the "stakeholders" who have an interest in the outcome of the decision
- Selection of a strategy on how the decision will be implemented.

Decision-making is the process through which alternatives are evaluated and a course of action is selected as the solution to a problem, opportunity, or issue.

Team decisions are inseparable from the team and the organizational planning elements: mission, objectives, goals, and strategies. Every principal decision

1 This section has been adapted from Cleland and Kerzner (1986).

Project Based Problem Solving and Decision Making: A Guide for Project Managers, First Edition. Harold Kerzner.

Companion Website: www.wiley.com/go/kerzner/projectbasedproblemsolving

made in the context of a project team's activities should help the team members to decide

- What is to be done?
- How will it be done?
- Who will do it?
- Why is it being done?
- When is it to be done?
- Where will it be done?

It is through the decision process that a determination of future action for the team is made. The decision-making process is a rational attempt to achieve results. Decision-making encompasses both an active and passive use of resources.

An active decision is one made through full consideration of the existing mission, objectives, goals, and strategies and a sincere attempt to make effective and efficient use of resources. A passive decision also consumes resources. If a decision-maker is not active in making a decision and procrastinates, organizational resources are consumed during the period of procrastination. If no deliberate decision is made, in reality, a decision has been made to do nothing. Resources continue to be consumed in the same manner that has been done in the past, often to the neglect of stakeholder consideration.

Stakeholders are the people, organizations, or institutions who contribute in some manner to or receive benefits from the decision. A product design team would have stakeholders composed of manufacturing, reliability, quality control, and design engineers plus representatives from customer service, marketing, and production control. Each of these professionals, as well as the organizations they represent, would have a vested and rightful interest in the decisions that are made in the design of the product. A decision made by the project team leader on behalf of the team should include stakeholder involvement. There is a valid reason for doing this. A decision must be implemented through a process of

- Delegation of appropriate authority and responsibility
- Allocation of human and nonhuman resources to support the decision
- Commitment of the people to make the decision workable
- Design a monitoring and control system such as the Earned Value Measurement System (EVMS) to determine if the decision is accomplishing the results that are desired.

Key decisions in modern organizations are influenced by laws, government regulations, environmentalists, and the actions or reactions of

- Competitors
- Suppliers
- Customers
- Collective bargaining units
- Employees

- Stockholders
- Creditors
- Local communities
- Professional societies.

Consequently, the decision-maker should always be mindful of how key decisions will be received by stakeholders who have a vested interest in the outcome of the decision to resolve a problem or take advantage of an opportunity.

2.1 Project Versus Business Problem-Solving and Decision-Making

Project managers today believe that each project that they are managing is part of the business and therefore they are managing part of a business rather than just a project. As such, project managers are expected to make business decisions as well as project decisions. However, there is a difference between project decision-making and business decision-making:

- Project decision-making focuses on meeting baselines, verification, and validation. Business decision-making focuses on market share, profitability, customer satisfaction, and repeat business.
- Project decision-making involves the project team whereas business decision-making may include marketing, sales, and senior management.
- There are multiple tools that project managers use for project decision-making but many of the tools used for business decision-making are mainly financial tools such as ROI, NPV, IRR, and cash flow analysis.
- The focus of project decision-making is project performance and deliverables whereas the focus of business decision-making is financial performance, business benefits, and business value created.
- The results of project decisions appear quickly whereas the results of business decisions may not appear for years.
- Most of the problems addressed as part of project decision-making are to maintain the baselines whereas business problems focus on alignment or changes to the business strategy.

2.2 Problem-Solving and Decision-Making in the Project Management Environment

In order to understand decision-making in a project management environment, it is first necessary to understand how the project management environment differs from the traditional environment we are all used to. The project management environment is an interaction between people, tools, processes, and routine business work that must be accomplished for the survival of the firm and project work. Project management activities may be considered secondary to the ongoing business activities necessary to keep the business ongoing. Because of the high level of risks on many projects and the fact that some of the best resources are assigned to ongoing business activities, the decision-making process on many projects can lead to suboptimal or ineffective decisions.

There are other factors that make decision-making quite complex in a project environment:

- The project manager may have limited or no authority at all to make the decisions even though they may have a serious impact on the project's outcome.
- The project manager does not have the authority to hire people to work on the project; they are assigned by functional managers after an often-lengthy negotiation process.
- The people on the project team may not be able to make meaningful contributions to the decision-making process. They may need to revert back to their functional managers for decisions to be made or approval of a decision.
- The project manager may not have the authority to remove poor workers from the project team without assistance from the functional managers.
- The project manager may not have any responsibilities for wage and salary administration for the project team members; this is accomplished by the functional managers. Therefore, the project manager may not be able to exert penalty power if the workers make poor decisions.
- The team members are most likely working on other projects as well as your project, and you have no authority to force them to work on your project in a timely manner.

Because of the project manager's limited authority, some project managers simply identify alternatives and recommendations. These are then brought to the executive levels of management, or the project sponsor, for the final decision to be made. However, some people argue that the project manager should have the authority to make those decisions that do not alter the deliverables of the project or require a change to the constraints and baselines.

2.3 The Impact of Constraints on Project Problem-Solving and Decision-Making

The boundaries on most projects are the constraints that are imposed upon the project team at the initiation of the project. For decades, project managers focused primarily on the constraints of time, cost, and scope when making decisions. But in today's environment, we are undertaking more complex projects and many of these have significantly more than just three constraints. Other constraints might include:

- Quality
- Safety
- Risks
- Impact on other projects
- Impact on ongoing or routine work
- Customer satisfaction
- Stakeholder satisfaction
- Capacity limitations
- Limitations on available resources
- Limitations on the quality of the available resources
- Compliance with government regulations
- Company image and reputation
- Cost of maintaining goodwill
- Maintaining ethical conduct
- Impact on the corporate culture.

Simply stated, all these constraints play havoc with the decision-making process. The time constraint probably has the greatest impact on decision-making. Time is not a luxury. The decision may have to be made even though the project manager has only partial information. Making decisions with complete information is usually not a luxury that the project team will possess. And to make matters worse, on some projects, we often have little knowledge on what the impact of the decisions will be.

2.4 The Impact of Assumptions on Project Problem-Solving and Decision-Making

At the beginning of a project, it is imperative for the project team to know not only what limitations or constraints are imposed upon them, but what assumptions have been made. The assumptions are related to the enterprise environmental factors that surround the project. Usually, the assumptions are listed in the project charter or business case, but more often than not, it is just a partial list.

To make matters worse, assumptions can and will change over the life of a project. The longer the project, the more likely it is that the assumptions will change. Good project managers establish metrics to track the assumptions to see if they have changed. Examples of assumptions that are likely to change over the duration of a project, especially on a long-term project, might include:

- The cost of borrowing money and financing the project will remain fixed.
- The procurement costs will not increase.

- The breakthrough in technology will take place as scheduled.
- The resources with the necessary skills will be available when needed.
- The marketplace will readily accept the product.
- Our competitors will not catch up to us quickly.
- The risks are low and can be easily mitigated.
- The political environment in the host country will not change.
- The parent company will not have a change in leadership.

The problem with having faulty assumptions is that they can lead to faulty conclusions, bad results, poor decision-making, and unhappy customers. The best defense against poor assumptions is good preparation at project initiation, including the development of risk mitigation strategies and possibly a structured approach for how complex decisions will be made and how the project team will manage assumptions that can change.

Assumptions have a major impact on the alternatives we select for the solution to a problem. Picking an alternative based upon faulty assumptions can have an undesirable impact on the project.

2.5 Understanding the Project Environment's Complexities

By now, you should have a reasonably clear picture of what decision-making and problem-solving are like in the project environment. Some important factors to consider include:

- There are numerous constraints imposed upon the project.
- The constraints can change in relative importance over the life of the project and new constraints can appear.
- The assumptions made at the beginning of the project may no longer be valid as the project continues.
- The project manager may not know all of the constraints even though some of them are listed in the project charter.
- The project manager most likely does not have a command of technology in the area where the problem exists.
- The project team members may likewise not have a command of technology in the area where the problem exists.
- The project manager and the team are expected to make a decision in a timely manner even though they may not have complete information.
- The client and perhaps all of the associated stakeholders may not be in agreement with the final decision.
- Expecting to always make the right decision is wishful thinking.

2.6 Selecting the Right Project Manager

Selecting the right project manager is essential. In general, the more complex the project, the more likely that problems will occur as expected, and the problems will require difficult decisions to be made. Unfortunately, not all project managers possess problem-solving and decision-making skills, and it may be impossible at the initiation of a project to identify that these skills may be critical for this project. Some project managers are excellent in project execution but poor at problem-solving. A history of experience on past projects seems to be the best way to select project managers that require these skills. These skills are not always easy to teach in a classroom.

Selecting a project manager with the right leadership style for certain projects is also essential. Some projects require the team to take risks, be creative, and be able to handle innovation. Unfortunately, the need for creativity and innovation may not be seen as a necessity at the onset of the project but may become a requirement when a problem occurs.

Some companies maintain a skills inventory database. At the end of each project, the project team is required to complete questionnaires that will be used to update the skills inventory database. The surveys include questions on creativity, problem-solving, innovation, and decision-making. The skills inventory database is then used to select individuals whose skills match the needs of the project assignment.

2.7 The Impact of the COVID-19 Pandemic on Project Management

In 2020, hundreds of thousands of people across the globe were affected by the COVID-19 pandemic. Public sector, private sector, and government leaders across all businesses had to respond quickly to the effects of the crisis, often with very little time to prepare. All of this was taking place in a rapidly changing VUCA (Volatility, Uncertainty, Complexity, and Ambiguity) environment characterized by a high degree of risk and uncertainty and accompanied by a lack of information for effective decision-making to take place.

Dealing with the crisis required that many projects be implemented quickly without having the luxury of sufficient time to use many of the traditional project management tools, techniques, and processes. The impact of the crisis affected project management leadership practices as well, especially the way that leadership should be applied and how decisions should be made. Most project teams had never been trained in how to deal with crises like this when executing both traditional and nontraditional projects. The information learned from managing projects during the pandemic has generated many best practices that are now being applied to all types of projects.

The pandemic has accelerated the recognition and need for significant changes in human behavior in project leadership. These characteristics are now seen as best practices and are becoming part of project management educational programs.

Trust

One of the biggest challenges in project management leadership is getting the team to trust that the decisions made by the project managers are based upon honesty, fairness, and ethical practices that consider the best interests of both the projects and the workers. Trust will encourage workers to become more engaged in the project and make it easier for the PMs to obtain their commitment. Trust can be destroyed if the PMs act in an unethical manner, abuse the power of their position, or demonstrate toxic emotions toward the team.

The pandemic placed workers under significantly higher levels of stress. Many workers quickly recognized the added complexities of their job with the reality of having to work from home. Working on virtual teams was a new experience for many workers. There was now a much heavier reliance on the project managers to provide guidance to the team members for their day-to-day activities as well as keeping them updated on how the company was planning to recover the business during the crisis, especially if there was a lockdown. The need for effective collaboration to establish trust is now seen as a best practice.

To build up and maintain trust, project management leadership had to provide emotional and interpersonal support for the team on a continuous basis. This was a new experience for many project managers who were just beginning to realize the importance of obtaining worker trust and its impact during a crisis.

Communications

Leadership during a crisis must focus on persuasion rather than formal authority and coercion. This is best achieved through effective communication. Active listening practices accompanied by emotional intelligence concepts and empathy should be used to understand workers' concerns.

When team members work from home for an extended amount of time, they rely upon media interaction and engagement to understand what is happening during the crisis. The greater the uncertainty during the media engagement, the greater the anxiety in the listeners.

Project managers must be willing to give up the idea that information is power and support the transparency of making information available. The information should provide a clear and realistic view of what is currently happening and possibly an optimistic view of future expectations. This is now becoming a best practice on all projects.

Decision-Making

On traditional projects, PMs often adopt a wait-and-see attitude before making decisions with the belief that all possible scenarios must be considered, including the worst-case scenario. Unfortunately, all of this takes time. During a crisis, delaying or not making a decision in a timely manner is generally a bad idea and can lead to a potentially worse outcome than expected by missing project opportunities. Decision delays can also create an environment where workers no longer have any credibility in the PM's leadership ability.

The pandemic has taught us that we must be willing to make project decisions rapidly based upon whatever information we have even if the information is imperfect or incomplete. PMs must do the best they can with the information available to anticipate what can go wrong.

During a crisis, strategic decisions are usually made for both the long-term and short-term benefits whereas project decisions may require a different focus. Some strategic decisions may be designed around the organization's core values which may not be in the best interests of the project's objectives. Other types of decisions may need to focus on the beliefs and values of the people affected by the decision. But regardless of the decisions made, the PM must spend sufficient time with all players, even if one-on-one in virtual meetings, to explain the meaning of the decision and what actions are expected from them.

Project Control Center

In traditional project management, with well-defined scope and expectations, critical decisions are most often made by the team in open meetings and then reviewed as needed with just the PM and either the sponsor or governance committee. However, during a crisis, executives often believe that they must take control resulting in the removal of leadership responsibility from the project manager and possibly the project sponsor. Meetings may be held behind closed doors with only a select few in attendance. These secretive meetings often do not include the PMs or project team members even though the resulting decisions may have a serious impact on the direction of the projects.

Using the hierarchical approach for making critical decisions during a crisis runs the risk of making bad decisions by overlooking critical information. The result is most often mistrust by the key players and a lack of commitment to the decisions.

Effective decision-making during a crisis should be done through brainstorming sessions. Having the right people in attendance, the greater the likelihood that critical information will not be overlooked, and more perspectives will be considered. Many of the participants may not have been part of the original project team if the crisis concerned just your project. The brainstorming sessions should include suppliers, distributors, and other strategic partners that may be experiencing the same risks and uncertainties as you and must understand the rationale for whatever decisions are made and how it impacts them. Their support may be essential.

There may be frequent brainstorming sessions and workers will want to know what decisions are being made and if they will be impacted. It may be beneficial to establish a project control center or project nerve center for the control of information to the workers.

Change Management

The result of crisis decisions frequently leads to change management activities as strategies based on new or promising ideas are designed for implementation. During a crisis, companies try to reduce costs and they often begin with abandoning educational activities. This can lead to detrimental results if the training and education are essential to fulfill and implement a strategy that requires organizational changes.

Conclusion

The magnitude and speed of crises are often unpredictable as we have seen from the COVID-19 pandemic. Many of the actions that companies have taken during a crisis are now becoming standard practices when managing many types of projects and are treated as best practices. This includes better ways to manage problem-solving and decision-making activities.

2.8 Differences Between Program and Project Problem-Solving and Decision-Making[2]

From the beginning of modern project management practices, there has been considerable confusion concerning the relationship between project and program management. The terms have often been used interchangeably. There have been articles written on the differences.[3] But now, partially due to the COVID-19 pandemic, organizations are looking much more closely at the differences and whether their limited funding should focus more on program rather than project

2 Adapted from Zeitoun and Kerzner (2022).
3 Weaver (2010).

management efforts. An important issue that needs to be considered is the understanding of the differences in problem-solving and decision-making.

The ***PMBOK® Guide*** provides the following definitions:[4]

- **Project:** A temporary endeavor undertaken to create a unique product, service, or result.
- **Program:** Related projects, subsidiary programs, and program activities that are managed in a coordinated manner to obtain benefits not available from managing them individually.

From a cursory position, projects focus on the creation of unique deliverables usually for a single customer. Programs focus on the synergistic opportunities that can be obtained from managing multiple projects to create business benefits and business value for both the organization and its customers.

There are textbooks that simply define program management as the management of multiple projects. However, there are other factors that can create significant differences between projects and programs. Managing a project as if it was a program can lead to significant cost overruns and cancellation. Managing a program as if it was a project can lead to significantly less than optimal results and failure.

There are numerous factors that can be used to differentiate programs from projects.[5] Some commonly used factors to identify the boundaries of projects and programs include:

- Type of products and services produced
- Industry type and characteristics
- Number and types of customers/stakeholders that will benefit
- Impact on business success
- Strategic risks
- Methodology used for implementation
- Size of the project or program
- Impact of environment factors such as the VUCA environment
- Complexity of the requirements
- Technology required and availability
- Strategic versus operational decision-making.

Perhaps the greatest difference is in the organizational behavior factors, the leadership style selected, the interaction with the team, and the decisions that must be made. When managing a small project, we make decisions that are usually in the best interest of the project. But when managing a program, a decision

4 Project Management Institute (2017).
5 See Zeitoun (2023).

that might appear in the best interest of one project may not be in the best interest of several of the other projects within the program. Therefore, decision-making in a program environment must consider the best interest of all of the projects within the program.

Reengineering Efforts and Change Management

Change is inevitable. Sooner or later, all companies undergo changes, some more often than others. The changes usually result from project successes and failures and can be small improvements to the organization's project management processes, forms, guidelines, checklists, and templates, or they can be major reengineering efforts that impact the organization's business model. The changes can also affect just one project or all the projects within a program.

There must exist a valid justification for decisions to make changes. Some companies expend countless dollars on changes and yet fail to achieve the desired results related to their strategic imperatives. Lack of employee buy-in is often a major cause of concern.

Project managers seldom take the lead in implementing reengineering efforts other than for the rare situation where the impact is on just one project. If the changes impact several projects that are not connected to a specific program, the leader of the change effort may be a steering committee. Program managers, on the other hand, must function as the change manager especially when the changes may impact many or all the projects within their program. They must expect more problem-solving and decision-making sessions than project managers.

Perhaps the most important responsibility of the person leading the change is communicating the business need for the change. Project managers may expect senior management or an executive steering committee to assume this role especially if the impact on their project is minimal. Program managers cannot and must not abdicate this responsibility to others especially if there could be a significant impact on the program's deliverables, customer and stakeholder expectations, and long-term financial considerations.

Face-to-face communication is essential, especially to get worker buy-in. Workers are always concerned as to how the changes will affect their job and whether new skills will be needed. People are fearful of being removed from their comfort zone especially if their vision of the future is uncertain. The impact of decision-making on programs is more likely to have a major impact on an employee's career.

Workers assigned to individual projects may have the option of requesting reassignment to those projects where they can remain within their comfort zone. Workers assigned to programs may be working on several projects within the program and reassignment may be impossible. Part of the program manager's face-to-face communications must include (i) the business need for the change,

(ii) how and when the change will take place, and (iii) his/her expectations of the workers after the change occurs. Workers are more likely to respond favorably to the changes if the information is provided by the program manager face-to-face rather than from someone not directly affiliated with the program.

Career Advancement Opportunities

Everybody seeks the opportunity for career advancement. In project management books and training courses, we stress that one of the roles of effective project management leadership is to help workers improve and to help meet their expectations while assigned to the project. Unfortunately, this is easier said than done in traditional project management. Some reasons for this include:

- PMs may have little or no authority over the workers and cannot hire or fire.
- PMs may not have the authority to conduct official performance reviews and may not be asked to provide recommendations to the functional managers on how well or poorly a worker performed.
- PMs may not possess the technical knowledge needed to evaluate worker performance.
- Workers may be assigned to multiple projects and the PM may not have sufficient time to evaluate worker performance.
- Project budgets may not have funding for the training workers need to advance their careers unless the training is specifically related to the project they are currently working on.

In most organizations, PMs do not conduct formal employee performance reviews. Worker performance improves when the worker exhibits his/her performance under the eyes of the person performing his/her review. If the worker on a project receives conflicting instructions from the project manager and his/her functional manager, the worker usual goes with the person conducting his/her performance review, which is usually the functional manager.

Program managers usually do not have all of the restrictions mentioned above. Workers may be assigned full-time on several projects within the program, and the program manager may be authorized to make a significant contribution to the worker's performance review process. Program managers generally get to know the workers on their programs better than the project managers.

Another important factor is the relationship between strategy and management/career development. Business strategies are formulated and executed through projects. Project management books and articles are now being written that identify the importance of aligning projects to business strategies. Why should executives authorize work on a project that is not aligned to one or more strategic business objectives?

There must exist a line-of-sight between senior management and project teams whereby team members understand the linkage to and importance of the strategic business objectives. Generally speaking, the linkage is often more evident between business strategy and programs than between business strategy and individual projects.

If a program has a long-term strategy and the workers recognize this strategy, then the program manager may be able to motivate the workers when they recognize the advancement opportunities through the business strategy for the projects within the program. Programs with growth strategies generally offer workers more management development opportunities than functional organizations that cater to single projects and focus on stability and possibly retrenchment strategies. Also, the reengineering efforts discussed previously, if explained carefully to the workers, may identify career opportunities.

Data-Driven Risk Management

An important lesson learned from the COVID-19 pandemic was that traditional approaches to risk management may be ineffective during a crisis. The risk associated with a failed project may be inconsequential compared to the failure of a long-term program.

Program managers require significantly more data than project managers for risk assessment since many of the projects within the program may be strategic rather than traditional or operational projects. Data-driven risk management will require access to information warehouses and business intelligence systems. As stated by the authors:[6]

> The changes that have been taking place in business and in the way of working of programs/projects have led to an unprecedented level of uncertainty that make the topic of estimating and the associated risks central to the success of the strategic initiatives... It is in our view that digitally enabled estimating requires innovation in order to create a commercially successful product, which also means that the team members must understand the knowledge needed in the commercialization life cycle starting from the early projects' stages.

Some project managers might simply walk away from a failing project without considering any decisions that could be made to rescue the effort and then move on to their next assignment. Program managers have significantly more at stake and focus on way to salvage as much business value as possible.

6 Kerzner and Zeitoun (2022).

Risk management is often looked at differently whether seen through the eyes of a project or program manager. Project managers tend to focus on negative risks, namely the likelihood of something bad happening and the resulting consequences. The intent is to reduce negative risk and the ways we have taught it according to the ***PMBOK® Guide*** are with strategies to avoid it, transfer it, mitigate it, or accept it.

Program managers must deal with positive risks as well as negative risks. Positive risks are opportunities to increase the business benefits and business value of the activities within the program based upon what is in the best interest of both the parent company as well as its customers. Opportunistic strategies include accepting it, exploiting it, transferring it, and enhancing it. Effective risk management activities for programs must consider both negative and positive risk management actions.

Stakeholder Relations Management

Stakeholder relations management is one of the knowledge areas in the ***PMBOK® Guide***. For the project manager, stakeholder relations management focuses heavily upon providing project performance feedback to stakeholders and engaging them in decisions and execution of the agreed-upon project plan. For program managers, the relationship is more complex.

Most project managers do not have the responsibility for marketing and selling of the deliverables of their projects. Project managers move on to their next assignment after project closure regardless of the business value of the deliverables and might never work with the same customers again. Program managers view themselves as managing a portfolio of projects with the strategic intent of creating a long and profitable life expectancy for the deliverables created within the program. Therefore, program managers may need much more sales and communication skills than project managers.

Both project and program managers have a vested interest in quality management, but often for different reasons. For the project manager, quality is most often aligned with the organization's Customer Relations Management (CRM) program which looks for ways to sell more of the existing deliverables to the customers in the short-term. The focus is on short-term thinking and quick profits.

Program managers have more of a long-term and strategic perspective by focusing on Customer Value Management (CVM) rather than CRM efforts. The intent of CVM is to get close to your customers to understand their perception and definition of value and what value characteristics will be important to them in the future. This allows the program managers to create meaningful strategic objectives aligned to their customer base. It provides a much closer and stronger relationship with important customers. This can lead to a lifetime bonding with critical customers and create a sustainable competitive advantage. CVM activities provide valuable knowledge for program managers for the resolution of program problems and the resulting decisions to be made.

Multi-Project Management and Innovation

Traditional project management focuses on managing a single project within the constraints of time, cost, scope, and quality. Innovation requirements are usually not included in most traditional or operational projects. The emphasis is usually on the use of existing knowledge in the creation of well-defined deliverables. Program managers are required to perform in a multi-project environment using many of the concepts required for effective portfolio management practices. Multi-project management gives program managers more opportunities than project managers working on a single traditional or operational project.

Some of the issues that program managers must address comes from answers to the following questions:

- How well are the projects within the program aligned to strategic objectives?
- Must any of the projects be canceled, consolidated, or replaced?
- Must any of the projects be accelerated or decelerated?
- Does the portfolio of projects need to be rebalanced?
- Can we verify that organizational value is being created?
- How well are the risks being mitigated?

Answering these questions requires the program manager to utilize business and strategic metrics for informed decision-making regarding strategic opportunities, new technologies that may be needed, and new products/services customers expect. Strategic opportunities have a strong linkage to innovation activities that address the customers' definition of value.

Previously, we stated that program managers have a greater interest in CVM than CRM and therefore must address innovation opportunities in the multi-project environment. Program managers must recognize that innovation-based strategies can become the key drivers to maintain or create a sustainable competitive advantage. By managing all the projects within the program in a coordinated manner whereby each project may be related to other projects, program managers are able to create significantly more business benefits and business value than traditional project managers. Obtaining this synergy requires significantly more metrics and some expertise in innovation practices necessary to support CVM activities.

2.9 Problem-Solving in Matrix Management Organizational Structures

In the past 40–50 years, there has been a great deal of literature published discussing the pros and cons of matrix management as well as step-by-step instructions on how to make it work well. What many people failed to realize was that matrix management during this time was heavily oriented around traditional projects rather than strategic projects. The way that problems are identified, and decisions are made, can be quite different when using matrix management for strategic projects.

Traditional projects start out with well-defined requirements, a business case, a statement of work, and possibly a detailed work breakdown structure listing all the work packages needed to be accomplished. Strategic projects may begin with just an idea and the requirements are elaborated as the work progresses. Strategic projects can differ from traditional projects by requiring:

- Different types of organizational resources
- Different information requirements for decision-making
- Additional resources, such as subject matter experts, may be needed during problem-solving and decision-making sessions
- More time than usually required for traditional projects
- A greater understanding of the VUCA environment and its impact on the projects (volatility, uncertainty, complexity, and ambiguity)
- Greater participation from the senior levels of management

Traditional projects in a matrix organization usually focus on deliverables for a specific customer. The goal is short-term profitability. Strategic projects may focus on the long-term needs of each senior manager and long-term corporate sustainability. These differences can change many of the core processes used in current matrix management practices, namely how decisions are made and who makes the decisions, how much risk team members are willing to accept when working with limited initial project requirements, information needs that are subject to possibly continuous changes, and a willingness to accept numerous changes in scope and direction. These differences due to the new types of projects are expected to cause significant changes to how we worked previously in a matrix organization.

Understand Matrix Management

Most companies today have a need to execute many projects, often at the same time, are organized in some form of a matrix structure. The matrix organizational structure is the layering of one or more new organizational forms on top of the existing vertical hierarchy to obtain horizontal as well as vertical workflow at the same time.

The intent is to build cross-functional project teams. The horizontal lines superimposed on top of the vertical hierarchy are usually referred to as projects and are removed from the organizational chart when the projects are completed.

The concept of matrix management and horizontal (or multidirectional) workflow became common practice in the aerospace industry beginning in the 1970s mainly because of technical reasons stemming from the need for innovation, new product development, and improved quality management practices. The challenge was the need to solve often complex technical problems that required the sharing of information between functional (i.e. vertical) units. In simple matrix structures, each horizontal line on the organizational chart represented a team or project and is headed up by a project or program manager whose prime responsibility is in a liaison role to coordinate information across functional units.

The resources assigned to project teams are owned by the functional managers and are assigned to the projects after negotiations between the project and functional managers. Once the assigned resources have completed their portion of the work, they return to their functional units. However, during the execution of the projects, team members most often are subjected to multiple boss reporting requirements where they must report to both the project manager and their respective functional manager simultaneously. The situation is further complicated if the team members are subjected to different sets of constraints and instructions they must adhere to.

Matrix structures tend to transform organizations from a vertical to hybrid organizational forms focusing on the execution of projects. The benefits of effective matrix management practices include the sharing of knowledge, coordination of multifunctional tasks, and decision-making involving several participants.

Matrix Structure Challenges

While on paper the matrix structure looks quite good, there are significant challenges facing the leader of project teams. The challenges include:

- The resources are owned by the functional units
- The functional managers may not assign the right resources to the projects
- The functional managers may have their own priority system which is different from the priority of the project and are unwilling to assign the resources when needed thus causing unwanted fluctuations in resource assignments
- The resources may be assigned to multiple projects, each with a different workload requirement
- The assigned resources may take instructions only from their functional managers rather than the project managers, or simply may favor functional managers when conflicting instructions are given

- The project managers may have no authority over the assigned resources, have no input into performance reviews of the resources, and cannot get resources removed from the projects without assistance from the functional unit managers
- The budgets and schedules for the projects may be determined prior to the start of the project and may be unrealistic
- Conflicts that arise on the projects may not be able to be resolved by the project manager alone
- Not all team members will see the project the same way and may have different ways of communicating, thus adding complexity to the communication and integration needs of the projects
- Team members may not like the leadership style used by the project manager and resist accepting instructions and advice
- There are often numerous delays in project decision-making because the individuals in power to make the decisions are not active members of the project team
- The corporate culture may be different and in conflict with the project team culture.

Over the past several decades, advances in matrix management practices have provided solutions and recommendations for continuous improvement. Most of the above-mentioned issues have become manageable. However, as stated previously, the growth in the need for more strategic projects has created new challenges for matrix management practices.

Competing Cultures

The most significant challenge with strategic projects will be resolving the issues with competing cultures. This can cause havoc with project teams especially if each culture has a different approach to problem-solving and decision-making. It has taken several decades for senior management to accept the need for cooperative cultures for traditional projects. Traditional projects, with well-defined upfront requirements, focus more so on short-term rather than long-term profitability and decision-making. Strategic projects have the potential to influence the future careers of organizational decision-makers and team members, thus making it difficult for them to surrender decision-making rights to others as is the case with traditional projects. In addition, there is the fear that there might be a dilution in the effectiveness of senior management without control from above over strategic projects.

On traditional projects, team members may be assigned part-time or for short periods of time. On strategic projects, there is a greater likelihood that team members may be full-time and for longer periods of time. There is also the chance that many of the team members have never worked with one another previously. As such, the impact arising from each team member's culture from their parent line functional organization must be known. Some cultures value risk-taking, a desire to speak one's mind, and a willingness to participate in decision-making, while others do not.

If you want team members to become engaged in the project and trust the people they must interface with, you must perform cultural research and discover their expectations as well as what might offend them. Developing trusting relationships is critical if you want team members to see how they can meaningfully contribute to the success of the project.

Part of cultural research must include the team members' perception of authority, specifically who is in charge and is empowered to make decisions. Some cultures sustain the belief that functional authority comes before project authority. This could create problems if team members are receiving conflicting instructions from the project manager and their functional managers.

Executive Involvement

As stated previously, executive involvement in strategic projects may be significantly different from traditional projects. Executives may have hidden agendas for certain projects that may have an impact on their careers or future desires. Their interest may appear in just certain projects rather than all of the strategic projects in the portfolio.

Executive involvement may be just a need to know or a desire to control the decision-making process. It would be beneficial if the project leaders could interview senior management and find out their interest in certain strategic projects and how they wish to be involved (i.e. which problems they wish to be involved with). Having an ambiguous chain of command in a matrix organization leads to frustration and often poor performance and poor decisions.

Active executive involvement with strategic projects may be a necessity. Many team members may be new to matrix management and executives can help selling the concept. Executives can also assist in maintaining the proper balance of power in the matrix between the vertical and horizontal lines. Finally, involvement in decisions on strategic projects may require participation by more than one executive.

Team Members

In matrix management, you may find it necessary to work with people you have never met and do not report to you. They come from different cultures and, because of the nature of strategic projects, may not even be sure that they fully understand what they are expected to do.

Matrix management can provide projects with the best talent in the company regardless of location of the personnel. However, getting them engaged and excited to work together as a cohesive unit requires that you establish an atmosphere of trust with the team members.

Effective communication is a necessity in all types of matrix organizations but may be the most important skill needed by the project managers when dealing with strategic projects that begin with just an idea rather than fully defined requirements as in traditional projects. Strategic projects may require team members to think and act differently.

For decision-making to be effective, team members must have a clear understanding of what this project means to company and how they might benefit personally from a successful result. You should ask them what their expectations might be from being assigned to this project. Even though the team members are not your direct reports, you may need to convince them that you will inform their superiors of their performance even though you may not be asked to do so. As stated by Kinor and Francis:[7]

> First, you must build a relationship with each individual. This is the best way to create trust and prove you have their best interests in mind. When you can establish these types of emotional connections, you'll forge strong and long-lasting working relationships. This might mean scheduling a time to talk one-on-one – in person if they're local or over the phone or via a video conference – at the start of the project. Use this time to discuss the project and ask for input on how each individual can bring his or her best work to the table.

Matrix management practices with traditional projects have made organizations stronger. Hopefully, the same success will occur with strategic projects. If organizations cannot overcome challenges with cultural differences, executive involvement, and new team member involvement in a timely manner when addressing strategic projects, serious issues can arise that might inhibit the organization's success and strategy execution.

2.10 The Impact of Methodologies on Problem-Solving

For several decades, companies relied heavily upon the processes in the ***PMBOK® Guide*** for problem-solving and decision-making practices, mainly on traditional projects. Many of the problems and solutions were common to several projects.

---

7 Kinor and Francis (2016).

As we started expanding the types of projects to include strategic requirements, innovation, R&D, and other business-related endeavors, companies began realizing that the waterfall methodology needed to be replaced by other flexible approaches such as Agile and Scrum. The new types of projects were accompanied by new categories of problems and solutions. Some flexible methodologies make it easier for project teams to identify problems and find solutions. In the future, project teams may select the methodology for a given project based upon the critical issues they expect that would need resolution.

Understanding Methodologies

A methodology is a set of principles that a company can tailor and reduce to a set of procedures and actions that can be applied to a specific situation or group of activities that have some degree of commonality. In a project environment, these principles might appear as a list of things to do and are often manifested in forms, guidelines, templates, and checklists. The principles may be structured to correspond to specific project life cycle phases, such as in a construction or a product development project.

For many years, the project management methodology (PMM) used by many companies provided for a waterfall[8] approach to accomplishing work because the various project phases are accomplished sequentially. The waterfall approach became the primary mechanism for the "command and control" of projects providing some degree of *standardization* in the execution of the work and *control* over the decision-making processes. However, this standardization and control came at a price limiting those instances as to when this methodology could be used effectively. Typical limitations included:

- **Type of Project:** Most methodologies that were either developed internally or purchased "off-the-shelf" assumed that the project's requirements were reasonably well-defined at the outset. As such, the project manager made tradeoffs primarily based on time and cost rather than scope. This limited the use of the PMM to traditional or operational projects that were reasonably well-understood at the project approval stage and had a limited number of unknowns. This also provides some standardization for problem-solving and decision-making efforts. Strategic projects, such as those involving innovation, where the end product, service, or result, was much more difficult to define upfront, could not be easily managed using the waterfall approach because of the large number of unknowns and the fact that the requirements (i.e. scope) could change, and sometimes frequently.

---

8 Waterfall is one of a few words used to describe a life cycle that generally follows a serial path. Other words used are "predictive," "serial," and "traditional."

- **Performance Tracking:** With reasonable knowledge about the project's requirements, performance tracking was accomplished mainly using the triple constraints of time, cost, and scope. Nontraditional or strategic projects had significantly more constraints that required monitoring and therefore used other tracking systems than those offered by the PMM. Simply stated, the traditional methodology had extremely limited flexibility – and value – when applied to projects that were not operational.
- **Risk Management:** Risk management is important on all types of projects. But on nontraditional or strategic projects, characterized by their high level of uncertainty and dynamic changes in requirements, many organizations found that the standard risk management practices included in traditional methodologies were insufficient for the type of risk assessment and mitigation practices found in such a fluid environment.
- **Problem-Solving and Decision-Making:** With similarities among the projects, there were typical types of problems needing resolution. Most of the time, the decisions made were similar to other projects and based upon history. But as new types of projects appeared, previous ways of handling issues were no longer the best approach.
- **Governance:** For traditional projects, governance was often provided by a single person acting as the sponsor (if there even was one assigned!). The methodology became the sponsor's primary vehicle for command and control and used with the mistaken belief that all decisions could be made by monitoring just the project's time, cost, and scope constraints.

The Faulty Conclusion

Organizations reached the faulty conclusion that a single methodology, a one-size-fits-all approach, would satisfy the needs of almost all their projects. This mindset worked well in many companies where it was applied to primarily traditional or operational projects. But on nontraditional projects the methodology failed, and in certain cases, in spectacular fashion.

As the one-size-fits-all approach became common practice, companies began capturing lessons learned and best practices with the intent of improving the singular methodology. Many of the best practices were related to problem-solving and decision-making. Project management was still being viewed as an approach for projects whose requirements were reasonably well-defined at the outset, having risks that could be easily identified, and executed by a rather rigid methodology that had limited flexibility.

Concurrent with the adoption and widespread use of the single methodology, strategic projects that included innovation, R&D, and entrepreneurship were being managed by functional managers who were often allowed to use their own approach for managing these projects rather than follow the one-size-fits-all methodology. Using innovation as an example, we know that there are several types of innovation projects each with different characteristics and requirements. Without employing a flexible or hybrid methodology, management was often at a loss as to the true status of these types of projects.[9] Part of the problem was that professionals working on innovation projects wanted the "freedom to be creative as they see fit" and therefore did not want to be handcuffed by having to follow any form of rigid methodology that could dictate how problems should be handled and decisions made.

The Project Management Landscape Changes

Companies began to realize the benefits of adopting formal project management practices from their own successes, the capturing of lessons learned and best practices, and published research data showing a link between project success and the adoption of project management best practices. Furthermore, companies were convinced that almost all activities and initiatives within the firm could now be regarded as a project and they were therefore managing their business by projects (also known as a project-based business).

As the one-size-fits-all methodology began to be applied to nontraditional or strategic projects, the weaknesses in the singular methodology became strikingly apparent. Strategic projects, especially those that involve innovation, may not be completely definable at project initiation, and as such the scope of work can change frequently during project execution. In fact, it is in the execution of the project that the requirements become clear. Also, governance of the project takes on a different form requiring significantly more involvement by the customer or business owner thus mandating a different form of project leadership.

Moreover, the traditional risk management approach used on operational projects appeared to be insufficient for strategic projects. As an example, strategic projects require a risk management approach that emphasizes VUCA analyses:

- Volatility
- Uncertainty
- Complexity
- Ambiguity.

Significantly more risks are found on strategic projects where the requirements can change rapidly to satisfy turbulent business needs. This became quite apparent on IT projects that focused heavily upon the traditional waterfall methodology

9 For additional information on the complexity of managing innovation projects and how they can be overcome, see Kerzner (2023).

that offered little flexibility to the project team to adjust the project's parameters based on changing requirements. The introduction of an agile approach implemented through any number of agile frameworks such as Scrum, solved some of the problems but created others.

Agile frameworks focused heavily upon better risk management activities but also required a great deal of collaboration with the business side of the company and not every business professional had the time or the inclination to devote the amount of time required for such collaboration. Every approach, methodology or framework comes with advantages and disadvantages.

The introduction of agile frameworks gave companies a choice between a rigid one-size-fits-all approach or a very flexible agile approach. To be sure, not all projects are perfect fits for an extremely rigid or flexible approach; many projects are middle-of-the-road projects that may fall in between rigid waterfall approaches and the more flexible agile frameworks. Projects that fall into this category often use hybrid life cycles— a combination of agile and waterfall—which can be used as a transition path to full agile implementation.

Selecting the Right Framework

Today, many practitioners strongly assert that a key role of the project manager is to decide which type of life cycle to use on a project (i.e. waterfall, agile, or hybrid) given its many characteristics. Others contend that new frameworks can be created from the best features of each approach and then applied to a project. What we do know with a reasonable degree of confidence is that new customizable frameworks that afford practitioners a great deal of flexibility are being used today and that as more organizations adopt the agile approach to work accomplishment, additional frameworks will be developed in the future.

For example, today we see organizations mixing various agile approaches such as Scrum, Kanban, and eXtreme Programming (XP). Many companies also use hybrid approaches as mentioned above combining agile and waterfall in various ways such as developing a product using agile methods but rolling out that product globally employing the waterfall approach. Some projects predominantly use the waterfall approach with some element of agile, and yet other projects use a predominantly agile approach with some element of the waterfall approach.

Deciding which approach or framework is best suited to a given project is a current challenge experienced by project managers in many, but not all, organizations. Some companies have not attempted to implement agile in any meaningful way and are still trying to solve all project issues with a "one-size-fits-all" approach. But the day is rapidly approaching where all project teams will be the given the choice of which framework to use. The choice of which framework to use could be based upon the number and types of problems expected and the solutions needed. We must never forget that the focus of our work on projects is on delivering value to our customer on a frequent basis. Whichever framework gets us there is the one we should be employing.

The decision regarding the best life cycle to use can be accomplished with checklists and questions that address characteristics of the project such as flexibility requirements, type of leadership needed, team skill levels needed, how decisions will be made, and the culture of the organization. The answers to the questions will then be pieced together to help decide which life cycle approach will be the most appropriate under the circumstances.

Be Careful What You Wish For

Selecting the right framework may seem like a relatively easy thing to do. However, as stated previously, all methodologies and frameworks come with disadvantages as well as advantages. Project teams must then "hope for the best" but "plan for the worst." They must understand what can go wrong and select an approach where execution issues can be readily resolved in a timely manner.

Some of the questions focusing on "What can go wrong?" that should be addressed before finalizing the approach to be taken include:

- Are the customer's expectations realistic?
- Will the needs of the project be evolving or known at the outset?
- Can the required work be broken down and managed using small work packages and sprints or is it an all-or-nothing approach?
- Will the customer and stakeholders provide the necessary support, and, in a timely manner?
- Will the customer and stakeholders be overbearing and try to manage the project themselves?
- How much documentation will be required?
- Will the project team possess the necessary communications, teamwork, and innovation/technical skills?
- Will the team members be able to commit the necessary time to the project?
- Is the type of contract (i.e. fixed price, cost reimbursable, cost sharing, etc.) well-suited for the framework selected?

Selecting a highly flexible approach may seem, at face value, to be the best way to go since mistakes and potential risks can be identified early allowing for faster corrective action thereby preventing disasters. But what many seem to fail to realize is that the greater the level of flexibility more layers of management and supervision may need to be in place.

Today, there are many approaches, methodologies, and frameworks available for project teams such as "Agile," "Waterfall," "Scrum," "Prince2," "Rapid Application Development," "Iterative," "Incremental," and the list goes on and on. In the future, we can expect the number of available methodologies and frameworks to increase significantly. Accordingly, some type of criteria must be established to select the best approach for a given project while considering the problems that may occur.

2.11 The Need for Problem-Solving Procedural Documentation

Project management is today the most significant development in organizational systems management. Project management is both a methodology and an organizational system designed to obtain more effective and efficient utilization of the company's resources of manpower, money, information/technology, equipment, facilities, and materials. In the management of multidisciplinary tasks, the effective utilization of resources requires that work be integrated across functional lines and that information as well as problem-solving and decision-making takes place both horizontally and vertically.

Multidimensional flow that requires problem-solving and decision-making cannot be achieved solely through rigid policies and procedures accompanied by formal authority. There must also exist strong interpersonal efforts and collaboration to accompany the documentation. Executives, project managers, functional managers, and functional specialists must be willing to interact as necessary without the struggle for power or status.

In a project environment, interpersonal skills are, to a large degree, communications. This is a challenge since we often must cross organizational boundaries and must deal with personnel from various functions who might have different backgrounds, interests, and organizational objectives. Furthermore, the project team can be a diverse group of people who have never worked together before but is organized just for the duration of the project. Yet another challenge is to cope with the frequent changes of personnel due to organizational as well as personal causes, thus necessitating some form of documentation.

People communicate in many ways. Often, communications get filtered and somewhat distorted. For many reasons, agreements in a project environment must be in writing. In some companies, project management believes in the law that only what is on paper is really important. It is not uncommon, especially in those organizations just beginning to use project management, to rely heavily upon rigid policies and procedures on how decisions should be made.

Other companies rely upon standards and documents such as the PMBOK® Guide for guidance. Although these documents can be extremely useful, their intended use can bring forth additional issues requiring significantly more decisions to be made than expected.[10] Documentation is needed, but the right documentation.

In addition to the need for effective interpersonal skills and collaboration, an important facet of any project management system is to provide the people in the organization with procedural guidelines on how to conduct many project-oriented activities, including problem-solving and decision-making, and how to communicate in such a multidimensional environment. The project management policies, procedures, forms, and guidelines can provide some of these tools for delineating the process, as well as a format for problem-solving and decision-making. The specific benefits of procedural documents, including forms and checklists, are that they help to:

- Provide guidelines and uniformity of execution
- Encourage use of the correct documentation
- Support clear and effective communications
- Standardize progress reporting practices
- Unify project teams and keep them engaged
- Provide a basis for analysis of performance
- Document agreements for future reference
- Refuel commitments
- Minimize paperwork
- Minimize conflicts and confusion
- Delineate work packages
- Bring new team members on board easily
- Capture best practices for future projects.

Done properly, procedural documentation will ease the pain for resolving issues that must involve both the performing and the customer organizations. This involvement creates new insight into the intricacies of a project and its management methods. It also leads to visibility of the project at various organizational levels, management involvement, and support. It is this involvement at all organizational levels that stimulates interest in the project and the desire for success and fosters a pervasive reach for excellence that unifies the project team. It leads to commitment toward establishing and reaching the desired project objectives and to a self-forcing management system where people want to work toward these established objectives.

10 See Appendix (A).

Procedural Documentation and Methodologies

Project management methodologies, whether rigid or flexible, provide guidance on how to execute projects. Without procedural documentation, decisions may be made by guesses rather than evidence and facts. One of the most important reasons for investing in procedural documentation for methodologies is to show customers and stakeholders how their projects will be managed. During competitive bidding practices, companies often market and identify in their proposals their methodologies and procedural documentation so that customers can follow the execution of the project.

Rigid methodologies, such as the one-size-fits-all approach, contain procedural documentation extracted from documents such as the ***PMBOK® Guide*** and ***Standards for Project Management***. The procedural documentation changes because of updates to the standards or the implementation of best practices. There does exist some degree of commonality among companies for many of the forms, guidelines, and templates that are part of the procedural documentation. For flexible methodologies, the degree of commonality is somewhat less.

Challenges

With all these benefits from using procedural documentation, management is still often reluctant to implement or fully support a formal project management system. Management concerns center often around four issues: overhead burden, start-up delays, stifled creativity, and reduced self-forcing control. First, the introduction of more organizational formality via policies, procedures, and forms might cost some money plus additional funding to support and maintain the system without recognizing the benefits. Second, the system is seen, especially by action-oriented managers, as causing undesirable start-up delays by requiring the putting of certain stakes into the ground, in terms of project definition, feasibility, and organization, before the detailed implementation can start. Third and fourth, the system is often perceived as stifling creativity and shifting project control from the responsible individual to an impersonal process that enforces the execution of a predefined number of procedural steps and forms without paying attention to the complexities and dynamics of the project and its possibly changing objectives. The comment of one project manager may be typical for many situations:

> My support personnel feels that we spend too much time planning a project upfront; it creates a very rigid environment that stifles innovation. The only purpose seems to be establishing a basis for controls against outdated measures and for punishment rather than help in case of a contingency.

This comment is echoed by many project managers. It also illustrates the potential misuse of formal project management systems to establish unrealistic controls and penalties for deviations from the program plan rather than to help to find solutions. Whether these fears are real or imaginary does not change the situation. It is the perceived coercion that leads to the rejection of the project management system. An additional concern is the lack of management involvement and funding to implement the project management system. Often the customer or sponsor organization must also be involved and agree with the process for planning and controlling the project.

How To Make It Work

Few companies have introduced project management procedures with ease. Most have experienced problems ranging from skepticism to sabotage of the procedural system, especially for problem-solving and decision-making. People feel they are now being restricted and removed from their comfort zone. Realistically, however, project managers do not have much of a choice, especially for the larger, more complex projects and programs.

Developing and implementing such a system incrementally is a multifaceted challenge to management. The problem is seldom to understand the techniques involved such as budgeting and scheduling, but to involve the project team in the process, to get their inputs, support, and commitment, and to establish a supportive environment. Furthermore, project personnel must believe that the policies and procedures of the project management system facilitate communication, are flexible and adaptive to the changing environment, and provide an early warning system through which project personnel obtains assistance rather than punishment in case of a contingency.

Although project managers have the right to establish their own policies and procedures, many companies have taken the route of designing project control forms that can be used uniformly on all projects to assist in the communications process. Some companies believe this is a necessity so that some standard way of solving problems will exist. Project control forms provide two vital purposes by establishing a common framework from which

- The project manager will communicate with executives, functional managers, functional employees, and clients.
- Executives and the project manager can make meaningful decisions concerning the allocation of resources and resolution of problems.

Success or failure of a project depends upon the ability of key personnel to have sufficient data for decision-making. Project management is often considered to be both an art and a science. It is an art because of the strong need for interpersonal skills, and the project planning and control forms attempt to convert part of the "art" into a science.

Many companies tend not to realize until too late the necessity of good procedural documentation. Today, most companies with mature project management structures maintain a separate functional unit, such as a project management office or project management center of excellence, for forms control. For most organizations, uniformity is a must.

Developing an effective project management system takes more than just a set of policies and procedures. It requires the integration of these guidelines and standards into the culture and value system of the organization. Management must lead the overall efforts and foster an environment conducive to teamwork. The greater the team spirit, trust, commitment, and quality of information exchange among team members, the more likely the team will develop effective decision-making processes, make individual and group commitments, focus on problem-solving and operate in a self-correcting control mode. These are the characteristics which will support and pervade the formal project management system and make it work. When understood and accepted by the team members, such a system provides the formal standards, guidelines, and measures needed to direct a project toward specific results within the given time and resource constraints.

Effective Project Control Practices

Effective planning and control techniques are helpful for any undertaking. They are essential, however, for the successful management of large or complex programs and projects given the number of issues that may need resolution. Management studies and consultants have pointed out repeatedly the strong correlation between quality planning and overall project performance. Quality of planning, however, means more than just the generation of paperwork. It requires the participation of the entire project team, including support departments, subcontractors, stakeholders, and top management. It leads to a realistic project plan plus involvement, commitment, and interest in the project itself. A look at the major benefits of proper planning shows that it makes everyone's job easier and more effective, because it:

- Provides a comprehensive roadmap of your project or program
- Provides a basis for setting objectives and goals
- Defines tasks and responsibilities
- Provides a basis for directing, measuring, and controlling the project
- Provides a basis for review and decisions
- Builds teams
- Minimizes paperwork

- Minimizes confusion and conflict
- Indicates where you are and where you are heading
- Leads to satisfactory program performance
- Helps managers at all levels to accomplish optimum results with available resources, capabilities, environment, and changing conditions
- Provides guidance on how issues should be resolved.

Done properly, the process of project planning and execution must involve both the performing and the customer organizations. This involvement creates new insight into the intricacies of a project and its management methods. It also leads to visibility of the project at various organizational levels, management involvement, and support. It is this involvement at all organizational levels that stimulate interest in the project and the desire for success and fosters a desire for excellence that unifies the project team. It leads to commitment to the establishment and attainment of the desired project objectives and to a self-forcing management system where people want to work toward these established objectives.

Discussion Questions

1. Why is it more difficult to make decisions in a project environment than in a functional organization?
2. What are the differences between active and passive decision-making and the impact on effective resources utilization?
3. What are some of the differences between project and business problem-solving and decision-making?
4. Are project decisions considered as more or less important than ongoing business decisions?
5. How do the constraints on a project impact problem-solving and decision-making?
6. How do the assumptions on a project impact problem-solving and decision-making?
7. How has the COVID-19 pandemic impacted problem-solving and decision-making?
8. Are there differences between project and program problem-solving and decision-making?
9. How can a project management methodology impact how decisions are made?
10. Why is procedural documentation for problem-solving becoming part of project management methodologies?

References

Cleland, D.I. and Kerzner, H. (1986). *Engineering Team Management*, 227–228. New York: Van Nostrand Reinhold.

Kerzner, H. and Zeitoun, A. (2022). The digitally enabled estimating enhancements: the great project management accelerator series. *PM World Journal* XI (V), July.

Kerzner, H. (2023). *Innovation Project Management*, 2e. Hoboken: John Wiley & Sons Publishers.

Kinor, L. and Francis, E. (2016). Navigating matrix management. *Leadership Excellence* 33 (2): 23–24.

Project Management Institute (2017). *A Guide to the Project Management Body of Knowledge (PMBOK® Guide)*, 6e, 715. Newtown Square, PA.

Weaver, P. (2010). Understanding Programs and Projects—Oh, There's a Difference! Paper presented at PMI® Global Congress 2010—Asia Pacific, Melbourne, Victoria, Australia. Newtown Square, PA: Project Management Institute.

Zeitoun A.l. and Kerzner, H. (2022). *Differences Between Programs and Projects*, white paper.

Zeitoun, A.l. (2023). *Program Management: Going Beyond Project Management to Enable Value-Driven Change*. Hoboken, N.J.: John Wiley Publishers.

3

Understanding the Problem

3.0 The Definition of a Problem

To understand problem-solving, we must first understand what is meant by a problem. A problem is a deviation between an actual and desired situation. It is an obstacle, impediment, difficulty, challenge, or any situation that invites resolution; the resolution of which is recognized as a solution or contribution toward a known purpose or goal. The problem could be to add something that is currently absent but desired, to remove something that is potentially bad, or to correct something that is not performing as expected. Therefore, problems can be formulated in a positive or negative manner. Problems are formulated in a positive manner if the problem is to determine how to take advantage of an opportunity.

We tend to identify alternatives as being good or bad choices. If the decision-maker has all the alternatives labeled as good or bad, then the job of the decision-maker or project manager would be easy. Unfortunately, a problem implies that there exists doubt or uncertainty, or else a problem would not exist. This uncertainty can happen on all projects and therefore makes it difficult to classify all alternatives as only good or bad.

Some problems must be resolved, whereas other problems may be able to be delayed and then resolved sometime in the future. Some problems are "good" problems because the objective is to take advantage of an opportunity.

3.1 The Time Needed to Identify a Problem

One of the critical issues with problem-solving, and often overlooked, is the determination of how much time should be spent on defining the problem. The time it takes to resolve a problem is often a measure of its complexity. Spending too little

Project Based Problem Solving and Decision Making: A Guide for Project Managers,
First Edition. Harold Kerzner.

Companion Website: www.wiley.com/go/kerzner/projectbasedproblemsolving

time on problem definition can result in a poor formulation of the problem, create confusion, and most likely lengthen the time needed to make effective decisions. On the other hand, spending too much time on problem definition can lead to unnecessary delays in the time needed for effective decision-making for a solution. Adding more people to the problem-solving session cannot always reduce the time it takes to resolve the issue. In some cases, we end up creating additional problems for ourselves.

Textbooks on project management seem to focus more so on decision-making efforts than on problem definition. Pressure is often placed on project teams for "speed." While well-defined problems usually allow for faster solutions to be found, project managers tend to quickly convene the project team to resolve an issue with the misbelief that a large project team can quickly develop an understanding of the problem, share knowledge, and agree upon a recommended solution. Without a clear understanding of the problem, rapid decision-making can lead to the wrong conclusions and possibly elongate the project.

Customers and stakeholders want problems to be resolved quickly and often have the faulty impression that the people assigned to their project possess all the information necessary to resolve issues. This is not always the case. Subject matter experts not assigned to this project may need to participate in problem solving and decision-making. Senior management and governance personnel also create issues by punishing project teams for taking too much time before identifying a recommended solution to a problem.

Companies must determine the proper balance between too little and too much time for problem definition. There are four factors that impact the time for problem definition. They are:

- Problem causes versus symptoms
- The risks associated with the problem
- The complexity of the problem
- Team members' knowledge and experience with this type of problem.

Project teams often focus on the symptoms or results of a problem rather than the true causes of the problem. As such, they may end up identifying the wrong problem for resolution.

The risk associated with the project, or a potential problem, may not be known early on by the people involved in the project. Team members may need support from their functional area in identifying the risks and finding solutions. Subject matter experts may need to participate.

The complexity of the problem may not be known until the team meets to discuss the problem and team members share their views and understanding of the issues. Some problems may seem quite simplistic at first, but then evolve into complex situations as related issues emerge.

Understanding the risks and complexity is most often based upon the team members' previous knowledge and experience with these types of situations. Sometimes, if the problems are severe and can damage the project, changes will be made to the resources assigned to the project. When team members have different definitions for a problem, conflicts can occur and elongate the time needed to find an effective solution. The more disagreements that exist, the more time is needed for problem definition understanding. Agreements on the definition of the problem can eliminate downstream rework and schedule delays.

Having the right amount of information for understanding the problem is ideal. However, there are situations where too much information is provided, and the result is information overload. Information overload can result in delayed decisions, making the wrong decision, and addressing the wrong problem.

Some people like to talk and provide additional or unnecessary information that may be unrelated to the problem at hand. Team members then become overwhelmed with unnecessary information. The additional information can bring forth new problems that may or may not be related to the original problem. Information overload can result in delays in decision-making or making the wrong decision.

3.2 Not All Problems Can Be Solved

Problems imply that some alternatives exist. Problems that have no alternatives are called open problems. Not all problems can be solved or should be solved. As an example, in R&D and new product development, it may take as many as 50–60 ideas to generate one commercially new product. The cost could be prohibitive to evaluate these many alternatives. Another common problem is in software development, where "gold-plating" the project with additional features that are unnecessary could have a serious impact on the end date and cost.

Some problems cannot be resolved without a breakthrough in technology. Companies that do not possess the technical skill or financial resources to undertake these breakthrough projects leave them as open projects. The same can also hold true for projects involving company image, reputation, and goodwill.

Finally, there are those projects that require compliance to Government regulations. For these projects, which are almost always very costly, all of the alternatives are often considered as poor choices. When forced to comply, we select the best of the worst. But, more often than not, we leave them as open problems until the very last minute, hoping the problem will be forgotten or disappear.

3.3 The Complexities of the Problems

Not all problems have the same degree of complexity. Problem complexity determines whether we should address the problem or leave it as an open problem. Some factors used to identify problem complexity include:

- Relative magnitude of the problem
- The cost to resolve the problem
- The availability of qualified resources to be part of the problem-solving team
- How well the problem is understood
- The amount of information available to solve the problem
- Whether we have partial or complete information on the problem
- The amount of remaining work on the project that may be impacted by the solution to the problem
- How the client will view the solution to the problem
- How the stakeholders will view the solution to the problem
- The impact that the solution (or failure of the solution) can have on the project team members' careers
- Whether the team will be motivated to find a solution to the problem
- Whether viable alternatives can be found for a solution.

3.4 Techniques for Problem Identification

It is impossible to solve a problem or take advantage of an opportunity without first being able to identify that a problem exists. Most people know that a problem exists when the project fails to meet the baselines. Waiting for the problem to surface may limit the time available for finding the best solution. Techniques available to the project team for problem identification include:

- Selecting the right metrics and key performance indicators
- Using templates and checklists that provide a series of questions that can be asked to identify present or future problems
- Effectively using team meetings
- Using walk-the-halls project management
- Monitoring the enterprise environmental factors for critical changes
- Working with people involved in customer relations management programs and customer value management programs
- Performing project health checks

- Listening to complaints
- Establishing a project suggestion box
- Performing risk management and establishing risk triggers
- Reading literature related to your industry.

Many of the specific techniques, and the accompanying tools, will be discussed in later sections.

3.5 Individual Problem-Solving Conducted in Secret

Companies encourage project team members to bring forth all problems quickly. The quicker the problem is exposed, the more time is available for finding a solution, more alternatives are usually available, and the greater the number of resources that can assist in the solution.

Unfortunately, some people simply do not want to identify the problem with the hope that they can resolve it by themselves before anyone finds out about it. This is true for people who may have been involved in creating the problem and fear that this may be held against them during performance reviews. Reasons for this include:

- Damage to one's reputation and image
- Damage to one's career
- Loss of employment
- Able to solve the problem using their own ideas rather than the ideas of others
- Dislike asking for help from others
- Distrust of the solution that others might choose
- Fear of antagonism from colleagues and team members
- Preference to working alone rather than in a team.

In such cases, people try to solve the problem by themselves, in secret, before anyone finds out about it. In reality, the problem is often hard to hide.

3.6 Team Problem-Solving Conducted in Secret

Sometimes the entire problem-solving team is in collusion in hiding the problem. Unfortunately, problem-solving sessions clearly identify that a problem exists, and this alone could make it difficult to hide a problem. It is easier for one person to try to solve a problem secretly than for an entire team.

Based upon the severity of the problem, the actual problem could be withheld from the client, the stakeholders, and even your own management, although the latter is certainly not a good idea. Sometimes people are not informed about a problem even after a solution is found and implemented. There are several reasons for wanting the problem to be resolved quietly:

- The client and/or the stakeholders may overreact to the problem and dictate the solution
- The client and/or stakeholders may overreact to the problem and remove financial support
- The client may cancel the project
- Problem resolution requires the discussion of proprietary or classified information
- Open identification of the problem may cause people to be fired
- Open identification of the problem may cause damage to your company's image and reputation
- Open identification of the problem can result in potential lawsuits
- The cause of the problem is unknown
- The problem can be resolved quickly without any impact on the competing constraints and the deliverables.

3.7 Decisions That Can Convert Failures into Successes[1]

All projects run the risk of failure. This includes traditional projects that have well-defined requirements based upon historical estimates as well as strategic projects such as innovation projects that may begin with just an idea. The greater the unknowns and uncertainties, as with many strategic projects, the greater the risk of failure.

When projects appear to be failing, there is a tendency for project teams to walk away from the project quickly in hopes of avoiding potential blame and trying to distance themselves from the failure. Project team members usually return to their functional areas for other assignments, and project managers move on to manage other projects. Companies most often have an abundance of ideas for new projects and seem to prefer to move forward without determining if failures can be turned into successes.

Part of the problem is with the expression that many project managers follow, namely, "Hope for the best, but plan for the worst." Planning for the worst usually means establishing project failure criteria as to when to exit the project and stop squandering resources. When the failure criteria point is reached, all project work tends to cease. Unfortunately, post-failure success may still be possible if the organization understands and implements the processes that can turn failure into success.

1 Adapted from Kerzner (2022).

There are many sources for failure. The most common cause of perceived failure is not meeting performance expectations. However, failure can also be the result of using the wrong processes, poor project management, and/or poor organizational leadership; making the wrong assumptions; having unrealistic expectations; having poor risk management practices; and focusing on the wrong strategic objectives. Many of these activities that were seen as the causes of failure can be overcome by reformulation of the project such that post-failure success may be attainable. Successful project management practices must focus on more than just creating deliverables. They must also focus on processes needed for converting failures into successes. Unfortunately, these practices are seldom taught in project management courses.

Sensemaking

Sensemaking is the process of making sense out of something that was novel, heavily based upon uncertainties or ambiguities and failed to meet expectations. Sensemaking is one of the commonly used techniques that may be able to convert failures into potential successes. Sensemaking cannot guarantee successful recovery but can improve the chances of success.

Turning failure into success requires, first, a culture of normalizing failures when innovating, and second, a process of problem formulation characterized by sensemaking that is both retrospective and prospective (Morais-Storz et al. 2020). Retrospective sensemaking addresses the issue of *what happened*, looks at the causes that led up to the potential failure, and tries to make sense out of the results. Prospective sensemaking addresses *what to do now* and envisions what the future might look like if we construct and implement a plausible new path (see Exhibit 3.1).

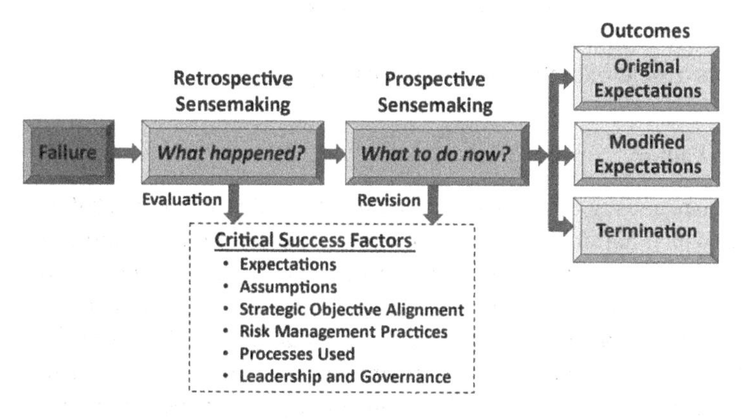

Exhibit 3.1 Post-failure Success Analysis

The leadership style that the project manager selects determines whether retrospective or prospective sensemaking will be emphasized as well as the approach to reformulating the project, if necessary. The outcome of sensemaking can be a reformulation of the original problem or the development of new plans that focus on somewhat different outcomes. Reformulation may be necessary if the team did not fully understand the way that the problem was initially presented to them, decision-making was being made based upon guesses rather than facts, or if significant changes occurred in the enterprise environmental factors. The knowledge gained from sensemaking may indicate that the original expectations are still valid, but that project must be reformulated. If the original expectations are no longer valid, then a new trajectory with perhaps modified expectations may be necessary.

The Need for New Metrics

Sensemaking requires more information than may be available in earned value measurement systems. When implementing traditional projects using the Waterfall Methodology and accompanied by well-defined requirements, decision-making is centered around time, cost, and scope metrics.

On strategic or innovation projects, given the high probability of failure, additional metrics may be required to determine the impact of variables that may have changed from the original problem formulation and resulted in a need for post-failure analysis. Some of the new metrics include:

- The number of new assumptions made over the project's life cycle
- The number of assumptions that changed over the project's life cycle
- Changes that occurred in the enterprise environmental factors
- The number of scope changes approved and denied
- The number of time, cost, and scope baseline revisions
- The effectiveness of project governance
- Changes in the risk level of the critical work packages.

Learning from Failure

Many project management educators advocate that the only true project failures are those from which nothing is learned. The cost for not examining failures can be very expensive if mistakes are repeated. Examination of a failed innovation project usually results in the discovery of at least some intellectual property that can be used elsewhere as well as processes that may or may not have been effective.

Organizations have looked at project failures for the purpose of capturing best practices so that mistakes are not repeated. Some people believe that more best practices can be found in failures than successes if people are not hesitant to discuss failures. Unfortunately, the best practices discovered from both success and failure analyses are usually related to changes in the forms, guidelines, templates, and checklists. Very little effort is usually expended at post-project reviews to identify behavioral best practices, which may very well have been the root cause of the failure.

Sensemaking allows organizations to address the psychological barriers that may have led up to the failure (Sitkin 1992). As stated by Morais-Storz et al. (2020):

> Failure is important for effective organizational learning and adaptation for several reasons. Failure helps organizations discover uncertainties, which are difficult to predict in advance (Sitkin 1992), creates learning readiness and motivates learning and adaptation (Cyert and March 1992), increases risk-seeking behavior (Kahneman and Tversky 1979), and act as a shock trigger to draw organizational attention to problems (Van de Ven et al. 2008).

Learning from failures can disrupt an organization if the outcome indicates that a significant change is needed such as in the organization's culture, business model, or processes. Innovation projects are generally not formulated with the same delicacy as traditional projects. The problem is further compounded by the fact that most innovation project managers are not brought on board the project until after organizational management has approved and possibly prioritized the project, identified the constraints, and stated the assumptions the team should follow. The innovation project manager, therefore, begins implementation with a project formulation that neither he/she nor the team participated in. Sensemaking usually occurs in a collaborative setting, perhaps involving most of the team members. Initial problem formulation, on the other hand, may be accomplished with just a few people, most of whom have never had to manage an innovation project.

The literature discusses various ways to formulate projects, but they are usually traditional or operations projects rather than innovation projects. There is a lack of literature on the link between failure and how the knowledge gained during retrospective sensemaking can benefit project reformulation on innovation-type projects by challenging complacency, the use of existing processes, and ineffective leadership in dealing with ambiguity.

The Failure of Success

Perhaps the greatest psychological barrier to strategic projects such as innovation is when an organization becomes complacent and refuses to challenge its assumptions, business model(s), and the way it conducts business. The organization is usually financially successful and believes that they will remain financially successful for years to come. As such, profitability and market share become more important than innovation, and this can create a significant psychological barrier for project teams. Eventually, the marketplace will change, profitability will be eroded, and strategic projects including innovation will become the top priorities. But by this time, the failure of success may have taken its toll on the company to a point where the company may never recover.

Characteristics of a potential failure of success environment include:

- Maintaining the status quo is essential
- Most decisions are made in favor of short-term profitability
- Maintaining the present market share is more important than investing in opportunities
- Senior management positions are filled by financial personnel rather than marketing and sales personnel, whose vision created favorable growth
- Executives refuse to challenge any of the business assumptions for fear of changing the status quo
- The guiding principles that led to success are not challenged
- The VUCA environment is expected to remain stable
- The company maintains a very low-risk appetite
- No changes are needed in the organization's business model
- The company will continue using the same suppliers and distributors
- No changes are necessary in organizational leadership
- No plans are made for future generations of managers
- The organizational culture is based upon command and control from the top floor of the building to the bottom and is not depersonalized to support innovation and free-thinking environment.
- No changes are needed in the organization's reward systems, which most likely are not based upon risk-taking
- Changes or continuous improvement efforts for project management processes are at a minimum
- There is not adequate funding for the needed innovation projects, other than perhaps small incremental innovation activities
- The failure of an innovation project is brushed aside if it has no immediate impact on profitability
- The use of sensemaking practices is nonexistent.

The critical issue facing innovation teams in successful companies is that all of the above characteristics create complex interrelations that team members must deal with. As stated by James O'Toole (1983):

> Innovation requires the ability to read changes in the environment and to create policies and strategies that will allow the organization to capitalize on the

> opportunities those changes create. Ironically, the most successful companies are likely to ignore environmental signals because it seems wildly risky to tamper when things are going well.
>
> Corporate reward systems also encourage behavior that is short-term, safe, and conservative. The 'punishments' for entrepreneurial failures, even when they are beyond a manager's control, are much greater than the rewards for successful risk-taking. Additional sources of discouragement for the innovator are the hurdles and delays associated with too many levels of approval needed for developing a new product or implementing a change in manufacturing processes or administrative practices.

Overcoming the failure of success may require companies to painstakingly change their management systems, culture, business models, and reward systems. Not all companies respond favorably to the risks associated with change management initiatives, even if the necessity exists. The greatest risk is when a change in the organization's business model may be needed, and people may be removed from their comfort zones. Some companies then perform innovation activities in a stealth mode for fear of upsetting the status quo.

A company that manufactured home products maintained an R&D group that maintained the status quo and focused heavily on incremental innovation rather than radical innovation efforts. Marketing wanted the R&D group to address several radical innovation projects. The R&D group assigned low priorities to the projects requested by marketing for fear of alienating senior management. Marketing then appointed workers from within the marketing organization to act as innovation project managers and perform the work in a stealth mode.

If an organization wishes to excel at innovation project management, they must understand the processes that can turn failures into successes. Discovery is essential. In addition, organizations must foster leadership styles that support failure analyses and accompanying change management efforts needed for retrospective and prospective sensemaking practices.

Discussion Questions

1. Why does it often take much longer than expected to resolve a project problem? What are the contributing factors?
2. Why do some project teams often rush into problem-solving sessions?
3. What factors make some project management problems difficult if not impossible to resolve?
4. How important are metrics for the identification of problems?
5. Why are some individual or team problem-solving sessions conducted in secret?

6. What is the primary reason why some project teams are now able to convert projects that appeared to have failed into successes?

References

Cyert, R.M. and March, J.G. (1992). *A Behavioral Theory of the Firm*. Cambridge, MA: Wiley-Blackwell.

Kahneman, D. and Tversky, A. (1979). Prospect theory: an analysis of decision under risk. *Econometrica* 47 (2): 263–291.

Kerzner, H. (2022). *Innovation Project Management*, 2e. Wiley: Hoboken. Chapter 11, Section: Post-Failure Success Analysis.

Morais-Storz, M., Nguyen, N., and Sætre, A.S. (2020). Post-failure success: sensemaking in problem representation reformulation. *Journal of Product Innovation Management* 37 (6): 483–505.

O'Toole, J. (1983). Declining innovation: the failure of success. *Human Resource Planning* 6 (3): 125–141.

Sitkin, S.B. (1992). Learning through failure: the strategy of small losses. *Research in Organizational Behavior* 14: 231–266.

de Ven, V., Andrew, H., Polley, D. et al. (2008). *The Innovation Journey*. Oxford: Oxford University Press.

4

Data Gathering

4.0 Understanding Data Gathering

Data-gathering techniques are a necessity in all Domain Areas of the ***PMBOK® Guide*** and, as such, can be considered as a crossover skill that is needed by everyone involved in the project. It may be difficult or even impossible to make the right decision without having all of the necessary information available. The larger and more complex the project, the greater the need for effective data-gathering skills. Unfortunately, the organizational process assets and enterprise project management methodologies and frameworks that project teams use do not necessarily provide all of the information needed to manage a project. The situation can become even worse if the team gathers the wrong information.

As an example of the use of data gathering, as part of the Scope Management area of knowledge in the ***PMBOK® Guide***, data-gathering techniques are used to collect the requirements, which is the process of documenting the stakeholders' needs to meet the project's objectives. Without effective data gathering, it may be impossible to meet the customer's and stakeholders' expectations.

When managing a project, the project manager must not expect that all of the information needed will be provided by the organization process assets. Although not often explicitly defined, it is the job of the project manager to gather all the necessary information whether it is done individually or through team members. Companies today maintain knowledge repositories where project teams can access much of the information they seek. Unfortunately, because of the uniqueness of most projects, the knowledge repositories may not contain all the information needed.

Project Based Problem Solving and Decision Making: A Guide for Project Managers,
First Edition. Harold Kerzner.

Companion Website: www.wiley.com/go/kerzner/projectbasedproblemsolving

4.1 Reasons for Data Gathering

In addition to collecting requirements, data-gathering techniques are used to:

- Decide what decisions to make
- What the impact of the decisions might be
- Decide what action items are necessary
- Determine the root cause of a problem
- Determine the causes for both favorable and unfavorable variances from a plan
- Determine the number and grade level of the resources that are needed
- Determine the risks that can occur and how they will be managed
- Select vendors
- Negotiate contracts.

While everyone strongly supports the use of metrics in project management, it should be understood that metrics are simply measurements against some standard or baseline. Metrics may identify part of the problem, but additional information must be discovered. Most metrics, in general, indicate performance rather than the cause of a potential problem. As an example, having an unfavorable cost variance may be seen as a problem, but it is the reason for the cost variance that is the true problem.

4.2 Data-Gathering Techniques

There are numerous techniques available for data gathering. The selection of the technique is based upon the information being sought out, the timing of the information, who will provide the information, the criticality of the information, and the type of decisions that the information must support. Each technique comes with strengths and weaknesses. Some data-gathering techniques can be done quickly. For the most part, data-gathering techniques are time-consuming. A partial list of techniques might include:

- Root cause analysis
- Panels of experts
- Facilitated work groups
- Questionnaires
- Surveys
- Interviews
- Observations and measurements
- Use of prototypes
- Diagramming techniques such as cause and effect diagrams
- Critical reviews of performance data
- Case study analysis.

Using just one technique may not suffice. It may be necessary to use several techniques to capture all of the required data.

4.3 Metrics and Early Warning Indicators

Most projects do not get into trouble overnight. By selecting the right metrics for a project, there are early warning signs that a problem is about to materialize. It is much easier to solve problems when they are small than when they become large.

Today, we believe that measurement techniques have advanced to the point where we can measure anything. Some measurement techniques and qualitative and others are quantitative. The days of having just time, cost, and scope metrics on a project are disappearing. Companies today have metric libraries that may contain hundreds of possible metrics.

At the beginning of a project, we perform risk management. Part of the risk management process is to identify potential problems that may occur on the project and to establish metrics and risk triggers that provide an early warning sign that a problem might occur. To assist us in doing this, we have lessons learned files, best practices and metrics libraries, and diaries on previous projects.

Establishing core metrics early in the project for many of the problems that might occur is certainly a good idea. However, it may not be practical. The more metrics you have, the more costly it becomes to track, measure, and report these metrics. But having some metrics is certainly better than having no metrics at all.

Having a metric that identifies a cost overrun or schedule slippage is nice to have but it does not identify the cause of the problem. But if you accompany this metric with another metric that identifies the number of assigned resources, or the quality (pay grade) of the assigned resources, you may have a better understanding of the cause of the problem or where to begin looking.

Experienced governance personnel often can predict what problems may occur on certain projects. They then establish metrics specific to these problems to help provide information for problem-solving.

4.4 Questions to Ask

Effective data gathering requires an understanding of what questions to ask. While it is true that the questions will be predicated upon the type of problem, typical questions might include:

- Are there any other resources of subject matter experts that can help us with this problem?
- How many problems do we have?

- Are there hidden problems that are below the surface?
- What is the extent of the problem?
- Is the problem getting worse, getting better, or stable?
- Did this problem exist previously on other projects?
- Can the problem be quantified?
- Can we determine the severity of the problem?
- What physical evidence exists to identify the problem?
- Who identified the problem?
- To whom was it first reported?
- Is there an action plan to collect additional information?
- Do we have the right team members addressing this problem?

4.5 Establishing Structure for Data Gathering, Problem-Solving, and Decision-Making

History has shown that project management performs significantly better when projects have a roadmap for all of the domain areas, namely initiation, planning, execution, monitoring and controlling, and closure. And of course, included in each of the domain areas are problem-solving and decision-making activities, all involving some form of data gathering. Most project managers prefer some sort of structure beginning with data gathering. The roadmaps need not be based upon rigid policies and procedures but can be constructed using forms, guidelines, templates, and checklists. The latter provides the project team with significantly more flexibility when managing the project.

More companies today are developing their own processes for data gathering, problem-solving, and decision-making. The processes are supported by templates which undergo improvements from captured lessons learned and best practices. There is a large amount of similarity between companies in the steps they use for problem-solving and decision-making. Without these templates, data gathering, problem-solving, and decision-making become ad hoc rather than structured processes.

4.6 Determining the Steps

Rational thinkers prefer an analytical approach to data gathering, problem-solving, and decision-making using sequential steps. There are several steps a company can choose from when setting up an approach:

- Recognizing the problem
- Understanding the problem
- Gathering the data

- Understanding the environmental impacts
- Understanding the assumptions
- Understanding the constraints
- Understanding the boundaries of the problem and the solution
- Convening the problem-solving team if not already done
- Generating alternatives
- Redefining the assumptions and constraints
- Evaluating the tradeoffs
- Evaluating the impact of the solution
- Selecting the best option
- Getting approval for the option
- Implementation of the alternative
- Monitoring and control of the solution.

Most companies perform all of these steps, but all of the steps may not be clearly defined as part of the company's approach. Also, many of the steps may be done in parallel rather than sequentially.

Discussion Questions

1. What are some of the risks associated with data gathering?
2. Do typical project metrics identify the root cause of a problem or performance?
3. If we believe that today we can measure just about anything, why do many project teams still focus on measuring just time, cost, and performance?
4. Should some sort of structure be established for data gathering on projects?
5. What steps in data gathering generally take the longest? Are the answers project specific?

5

Meetings

5.0 Problem Analysis Characteristics

When discussing meetings, it is important to differentiate between problem-analysis and decision-making meetings. The concepts are completely separate from one another. Problem analysis must be done first, and then the information gathered in that process may be used toward decision-making.

Problem analysis involves:

- Analyze performance; what should the results be against what they actually are
- Problems are merely deviations from performance standards, baselines, or expectations
- Problems must be precisely identified and described
- Problems are caused by some change or variation from a distinctive feature
- Something can always be used to distinguish between what has and has not been affected by a cause
- Causes to problems can be deduced from relevant changes found in analyzing the problem
- Most likely cause to a problem is the one that exactly explains all the facts.

There are tools available to assist in identifying the problems. Root cause analysis can be viewed as such a tool as can the earned value measurement system if used properly. Once again, the constraints imposed on the project may dictate the amount of time available for problem analysis. The potential risks associated with the problem may also dictate the amount of time and funding spent on problem analysis and decision-making.

Project Based Problem Solving and Decision Making: A Guide for Project Managers,
First Edition. Harold Kerzner.

Companion Website: www.wiley.com/go/kerzner/projectbasedproblemsolving

5.1 Real Problems Versus Personality Problems

We generally believe that most problems are real and need to be resolved. But that is not always the case. Some problems are created based upon the personalities of the individuals and possibly hidden agendas for certain people. Some people create problems unnecessarily as long as they can somehow benefit, perhaps by being the only person capable of solving the problem. Some examples include:

- Resolution of the problem will get you more power.
- Resolution of the problem will get you more authority.
- Resolution of the problem will diminish the power and authority of others.
- You are the only one with the capability to resolve the problem and it will improve your image and reputation.
- You will be regarded as a creating thinker.
- It will look good on your resume.
- It will look good during performance reviews.
- It will guarantee you long-term employment.

It is important to discover first of all whether or not the problem is real and if a simple solution exists for it. Many years ago, a department manager was afraid that downsizing would take place and that he would lose his position as a department manager. To protect himself, he gave the workers assigned to project team conflicting instructions knowing that problems would occur and that several projects might suffer. This resulted in rework and created problems for several projects. The department manager then called a problem-solving meeting with those project managers that were falling behind on their projects. The department manager then stated in a meeting with senior management that almost all of his employees were poor workers that needed constant supervision, and that the problem will be resolved by the department manager. He would provide these poor workers with much closer supervision. The department manager provided alternatives to the fictitious problem and stated that he would have the problems resolved within few months.

The department manager believed that his position was now secure. But the project managers were not fooled. The project managers found out the truth, and eventually, the department manager was fired for what he did. The project managers discovered that this was not a real problem that needed to be resolved using problem-solving and decision-making techniques.

Hidden agendas can be difficult to determine. The can exists with customers and stakeholders as well as with individuals within your company.

5.2 Determining Who Should Attend the Problem-Solving Meeting

Problems are not resolved in a vacuum. Meetings are needed and the hard part is to determine who should attend. If people are not involved in the problem or the problem is unrelated to the work they do, then having them attend these meetings may be a waste of their time. This holds true for some of the team members as well. As an example, if the problem is with procurement, then it may not be necessary for the drafting personnel to be in attendance. However, the exception might be that some people can come up with good ideas for a solution to a problem even though they are not impacted by the decision.

For simplicity's sake, we shall consider just two types of meetings: problem-solving and decision-making. The purpose of the problem-solving meeting is to obtain a clear understanding of the problem, collect the necessary data, and develop a list of workable alternatives accompanied by recommendations. More than one meeting will probably be required.

Sending out an agenda is important. The agenda should include a problem statement which clearly explains why the meeting is being called. If people know about the problem in advance, they will have a chance to think about the problem and bring the necessary information, thus reducing some of the time needed for data gathering. It is also possible that the information gathered will identify that the real problem is quite different from what was considered to be the problem at first.

It is essential that subject matter experts familiar with the problem be in attendance. These subject matter experts may not be part of the original project team but may be brought in just to resolve this problem. The subject matter experts may also be contractors hired to assist with the problem.

The people brought in for the identification of the problem and data gathering usually remain for the development of the alternatives. But there are situations where additional people may participate just for the consideration of alternatives.

5.3 Determining Who Should Attend the Decision-Making Meeting

The decision-making meeting is different from the problem-solving meeting. In general, all of the participants that were involved in the problem-solving meeting will most likely be in attendance in the decision-making meeting but there may be a significant number of other participants. Project team members should have the ability to resolve problems but not all of the team members have the authority to make decisions for their functional units. It is normally a good idea at the initiation of the project for the project manager to determine which team members possess this authority and which do not. Team members that do not possess decision-making authority will still be allowed to attend the decision-making sessions but may need to be accompanied by their respective functional managers when decisions are required, and voting takes place.

Stakeholder attendance is virtually mandatory at the decision-making meetings. The people making the decisions must have the authority to commit resources to the solution of the problem. The commitment could involve additional funding or the assignment of subject matter experts and higher-pay-grade employees.

Project managers are responsible for the implementation of the solution. Therefore, the project manager must have the authority to obtain the resources needed for a timely solution to the problem.

5.4 Creating a Framework for the Meeting

For the problem-solving meeting, it is important to create a mental framework of the problem beforehand, including what should be accomplished in the meeting, and the limitations. Not all of the people that will attend the problem-solving meeting will be familiar with the problem. Some may have just a cursory understanding of the problem and others may not have known that the problem even existed prior to the meeting.

The mental framework should include all of the information known thus far about the problem. Additional information will most likely be forthcoming in the meeting. If possible, the framework should be included in the invitation for the meeting and/or the agenda. Informing people about the problem prior to the meeting will get them to think about it and possibly even perform some preliminary research prior to the meeting. When people understand the framework prior to the meeting, they usually come to the meeting better prepared and may even recommend to the organizer of the meeting other people that should be invited.

5.5 Setting Limits on Problem-Solving and Decision-Making

Problem-solving and decision-making can go one for a long period of time. Limitations must be established early on. Some of the limits include:

- How much time is available to address the issue
- How much money the project is willing to commit to resolve the problem

- How many resources can be assigned to resolve the problem
- Whether the assigned resources will have the required skills
- Which facilities can be used for testing or other activities
- The importance of the project to the company
- The importance of the project to the client
- The importance of the project to the stakeholders
- The criticality of the project and the problem.

It is not uncommon for the limitations to be identified in the problem statement or even in the agenda. When people understand the limitation early on, decisions are usually made in a timely manner.

5.6 Identifying Boundary Conditions

The limitations discussed previously were limitations on the meeting. Limitations or boundary conditions must also be established for the solution to the problem, and the limitations can impact the alternatives that will be selected.

We know for a fact that, when solving a problem and identifying alternatives, we do not have an infinite sum of money to implement a solution or as much time as we want. We consider these as constraints, but they are also boundary conditions. Boundary conditions can be established by the client, stakeholders, and the eventual users of the deliverables. A partial list of boundary conditions might be:

- Staying within the project's constraints on time, cost, quality and scope
- Without increasing the risks on the remaining work on the project
- Without altering the scope of the remaining work on the project
- Without altering the company's normal flow of work
- Without gold-plating the solution
- Without including unnecessary functionality
- Knowing that only a limited number of additional resources are available for solving the problem and implementing the solution
- Without violating Regulatory Agency requirements such as those established by OSHA and EPA
- Without driving up the selling price of the product beyond what the customers will pay for it.

5.7 Understanding How People React in Meetings

Team meetings that are involved in problem-solving and decision-making often get people to act in an irrational manner especially if the outcome of the meeting can have a negative impact on them personally. This is particularly true for people that are closely identified with the cause of the problem. You may also be inviting people you have never worked with previously and you have no idea how they will react to the problem or the solution. Attitudes that must be closely watched in these meetings are:

- **The aggressor:** criticizes everyone as being part of the problem even if the cause is just one person
- **The devil's advocate:** always argues that there are other causes of the problem and refuses to become a believer in the real cause unless threatened
- **The dominator:** tries to take over the meeting and professes to know everything about the problem and what the solution should be. This is seen as a chance for glory.
- **The recognition seeker:** always argues in favor of his/her impression of the problem and his/her solution to the problem
- **The withdrawer:** may be afraid of criticism and does not want to be identified as being part of the problem.

There are other attitudes that can appear in meetings, but these are the ones that seem to occur most frequently in project management.

5.8 Working with Participants During the Meetings

If problems must be identified and alternatives must be found, then working with people in the traditional manner may not work. This is particularly true if you have people in the meeting with the attitudes discussed previously. In these types of meetings, there will be bickering and conflicts. The project manager or meeting leader must create an environment that will lead to a successful outcome. Given the makeup of the participants and the severity of the problem, there are things that the project manager can say or do that would make it easier for the people to participate. Some of the expressions that the project manager can use include:

- Is there a chance that this might be the real cause or is there a chance that this might work?
- Have we tried anything like this before?
- Do we know any other time when this has happened?
- Do we know of any other companies that have had similar problems?
- Your idea has a lot of merit!
- Your idea is great, but we may have to make a small change.
- What you said will really help us.
- Are we saying that . . .
- Let me say in my own words what I think you just said.
- Let's see if we can put this in perspective.

- Your idea and my idea are close together.
- Aren't we saying the same thing?
- Let's see if we are in agreement.
- Let's see how the rest of the team feels about this.
- Who hasn't given us their opinion yet?
- Are we prepared to make a decision? Or is there additional information we need?
- Should we keep our options open?

It should be obvious in most of these expressions that the project manager is trying to solicit feedback.

Based upon the severity of the problem to be resolved, the chairperson of the meeting may be someone other than the project manager. The chairperson may be someone specifically trained in facilitation skills. The project manager would attend the meeting but not function as the chairperson.

5.9 Leadership Techniques During Meetings

There are several techniques that the leader or the project manager can adopt to encourage the best possible outcome from the meetings. These include:

- Encouraging people to speak
- Asking probing questions
- Avoiding questions that may be counterproductive
- Keeping emotions under control
- Soliciting feedback
- Providing constructive feedback rather than personal criticism
- Understanding the people on the team and their needs and interests
- Understanding the legal implications of all of the alternatives and decisions
- Resisting attempts at gamesmanship.

5.10 Handling Problem-Solving and Decision-Making Conflicts

Conflicts and conflict resolution are a way of life in project management. Some conflicts may have a higher intensity level and are therefore more difficult to resolve. Not all conflicts are bad. People often argue in favor of their own opinion and if they continue to bring forth additional facts that may be important, then the conflict may be allowed to continue. These types of conflicts are often seen as "constructive" conflicts.

During problem-solving and decision-making sessions, it is often desirable to invite people that you know are opponents to the likely decision. These people generally bring forth a significant amount of data to support their position and quite often the additional information causes a change to the alternatives selected.

Not everyone will be in agreement with the source of the problem, the alternatives, or the decision. There is a chance that some of the conflicts may linger on long after the final decision is made. Expecting everyone to agree with the final decision is wishful thinking.

In an event, the project manager or meeting chairperson must have reasonable knowledge of conflict management. If a great deal of conflict is expected at the meetings, it may be desirable to have a professional facilitator act as the chairperson rather than the project manager.

5.11 Continuous Solutions Versus Enhancement Project Solutions

Not all problems need to be resolved immediately. Some problems may be needed to be resolved immediately whereas other problems can be clustered together and resolved later using enhancement projects. An example might be a software project that is designed to support inventory management at a company's manufacturing plants. After the project was well underway, the plants wanted changes to be made to the software. Resolving all of the issues as they came up was deemed unrealistic by the project manager because the launch date of the original project would be delayed continuously. In this case, the project manager was able to establish an enhancement project that would address many of these issues as a separate project that would be completed sometime after the original project was completed. Not all problems can be pushed into the future for resolution.

Not only can there be a schedule slippage with continuous problem-solving, but there may be a lack of qualified resources. Most companies do not have excess resources sitting idly waiting for an assignment. Unavailable resources can further elongate the schedule.

5.12 Problem-Solving Versus Scope Creep

Too often, project managers believe that most problems must be resolved using scope changes. The result is scope creep. Scope creep is the continuous enhancement of the project's requirements as the project's deliverables are being developed. Scope creep is viewed as the growth in the project's scope and is often rationalized as a necessity to solve all of the problems that come up when managing a project.

Scope creep is a natural occurrence, and we must accept the fact that it will happen. Scope creep can lead to beneficial results. But, on the other hand, not all problems require scope changes. We can argue that scope creep isn't just allowing the scope to change, but an indication of how well we resolve problems. In this regard, project managers should ask themselves:

- Do we need scope changes to resolve this problem?
- Is the customer always right regarding their view of the solution?
- Have we acted as the devil's advocate to see what would happen without a scope change?
- Will a solution that requires a scope change lead to other scope changes?

5.13 Problem-Solving and Decision-Making During Crisis Projects

Crisis problem-solving and decision-making are significantly different from ordinary project problem-solving and decision-making. Crisis projects are those projects that either have already or may result in the loss of life or possibly significant damage to the company. Examples include loss of life resulting from product tampering and the use of faulty equipment. Some of the major differences in problem-solving and decision-making when human life is involved include:

- Time is an extremely critical constraint rather than just an ordinary constraint
- Life cycle phases are measured in hours or days rather than weeks or months
- The chairperson of the problem-solving and decision-making teams may be the project sponsor or someone from senior management levels rather than the project manager
- Sufficient time may not exist for a complete understanding of the root cause of the problem
- The problem-solving and decision-making meetings may include representation from all of the stakeholders, including Government agencies, even those that were mainly observers in the past
- The news media may be actively interested in what you are doing to solve the problem and extreme care must be taken on how to communicate with them
- Communications with the media will normally be handled by someone from senior management rather than the project manager
- The alternative selected to resolve the problem must be able to be implemented as quickly as possible
- Future risks and the potential of future lawsuits may be an important factor in selecting alternatives.

5.14 Presenting Your Decision to the Customer

Solving a potential problem and making a decision on a solution does not mean that the decision will be implemented. You must still make a presentation to the client to get their buy-in on your recommendation. How you make the presentation is critical to getting acceptance for the decision.

Selling your recommendation must satisfy two requirements. First, you must show that your recommendation is technically, financially, organizationally, and legally an acceptable solution to satisfy your client's needs. Second, your client must understand the basis for the recommendation and fully agree that it satisfies their needs.

What you present and how you present it is critical. Rarely does a project team come up with a perfect solution to a problem. It may be impossible or unrealistic to meet all the constraints imposed on the problem. You may find it necessary to discuss many of the alternatives you considered and your rationale for choosing the alternative you are recommending. If this is not done, the client may have a concern that you did not consider all the alternatives and better approaches might still exist. If the client is still unhappy with your recommendation, you might also run the risk that the client will dictate the solution, and this might not be in your company's best interest or the best interest of the project.

The purpose of the briefing is to present the complete recommendation to an individual or large group to gain a consensus. Knowing who will be in the audience and their knowledge of the project helps you to properly design the presentation and anticipate questions that may be asked. It also helps you to consider how you will respond to any open issues that might still exist.

Knowing your audience helps you determine the proper wording for the presentation. As stated by King:[1]

> The style of the briefing will have to be structured to match the needs of this audience on this subject. However, there are some general guidelines for any presentation. Structure your presentation and your style based on the assumption that the audience is intelligent but uninformed. Structuring a presentation with this in mind should discourage a tendency to talk "down" to or "over the head" of the audience.

There are other factors that need to be considered in addition to developing a good presentation. According to King, other types of considerations that need to be addressed are:[2]

- The physical environment available for the presentation. This would include the size, shape, and location of the presentation room needed or the one that is available.

1 King (1981).
2 King (1981).

- The physiological mood of the audience. The number of friendly, unfriendly, or skeptical persons in the audience can have a significant impact on the way the briefing is structured.
- The length of the presentation. How much time do you need during the briefing to present the material to make the case that you would like to make? How much time can the audience devote to this subject?
- Relationship to other presentations. Are there other briefings or meetings on the schedule of this audience that may influence the frame of mind of the participants?
- Time of day. What time during the day would be the most appropriate time to present this subject to the audience? What time during the day has been scheduled for this presentation? Can the time be modified if it is desired?

Discussion Questions

1. How does a project manager determine if a problem is a "real" problem requiring resolution?
2. How easy or difficult is it to identify if the cause of a problem is the result of someone's hidden agenda?
3. What are the benefits of preparing agendas for meetings?
4. Who determines the attendees for a decision-making meeting?
5. Why should time limits be set for meetings?
6. What are examples of "boundary conditions" established for meetings?
7. If some meeting participants are in conflict over issues being discussed in a meeting, what should the project manager say or do?
8. Is there a relationship, if any, between problem-solving and scope creep?
9. How does crisis problem-solving and decision-making differ from traditional project problem-solving and decision-making?
10. Who should present the results of a meeting known to the customers?

Reference

L. Thomas King (1981). *Problem Solving in a Project Environment.* Wiley, p. 137.

6

Developing Alternatives

6.0 Finding Alternatives

A major part of problem-solving and ultimately decision-making involves the identification and analysis of a finite set of alternatives described in terms of some evaluative criteria. These criteria may be benefit or cost in nature, or the criteria could simply be the adherence to the cost, schedule, and scope baselines of the project. Then the problem might be to rank these alternatives in terms of how attractive they are to the decision-maker(s) when all the criteria are considered simultaneously. Another goal might be to just find the best alternative or to determine the relative priority of each alternative.

The number of alternatives is often limited by the constraints imposed upon the project. For example, if the actual schedule is exceeding the baseline schedule, then the project manager may have five alternatives: overtime, performing some work in parallel rather than in series, adding more resources to the project, outsourcing some of the work to a lower cost supplier or reducing the scope of the project. Each alternative will be accompanied by advantages and disadvantages. If the goal is to lower the costs, then there may be only one viable alternative, namely reducing the scope.

Because of the complexity of projects, the project manager cannot be expected to determine all of the alternatives in a vacuum. The team should be involved in the identification and prioritization of the alternatives. If the team does not have the expertise to determine the alternatives, then the project manager may need additional support from the subject matter experts in the functional areas. There are also situations where the stakeholders or external contractors may be able to provide the necessary information for determining alternatives.

Project Based Problem Solving and Decision Making: A Guide for Project Managers,
First Edition. Harold Kerzner.

Companion Website: www.wiley.com/go/kerzner/projectbasedproblemsolving

6.1 Variables to Consider During Alternative Analyses

There are several variables that must be considered when identifying and selecting alternatives. The variables are usually project-specific and based upon the size, nature, and complexity of the problem. However, we can identify a core list of variables that usually apply to the identification and evaluation of most alternatives:

- **Cost:** there is a cost associated with each alternative. This includes not only the cost of implementing the alternative, but the financial impact on the remaining work on the project.
- **Schedule:** implementing an alternative takes time. If the implementation time is too long or cannot be done in parallel with the other project work, then there may be a significant impact on the end date of the project.
- **Quality:** care must be taken that the speed to resolve a problem does not result in a degradation of quality in the project's deliverables.
- **Resources:** implementing a solution requires resources. The problem is that the people needed with the necessary skills may not be available.
- **Feasibility:** some alternatives may seem plausible on paper but may be unfeasible when needed to be implemented. Feasibility or complexity of the alternative must be considered. Otherwise, you could make the problem worse.
- **Risks:** some alternatives expose the company to increased risks. These may be future risks (or even opportunities) that will appear well after the project is completed.

6.2 Understanding the Features That Are Part of the Alternatives

Previously, we discussed some of the variables that must be considered looking at alternatives. Now, we look a little deeper into the features that are part of the alternatives. Many times, there are several features that can be included in each of the alternatives. Part of understanding the boundary conditions is to know the importance of each feature. The features can be classified as:

- **Must have:** any alternative that does not include this feature should be discarded.
- **Should have:** these are features that in most situations should be included in the alternatives that are being considered. Failure to consider these could result in a degradation of performance. Some of these features may be omitted if including

them results in unfavorable consequences when trying to satisfy the competing constraints.

- **Might have:** these are usually add-ons to enhance performance but not necessarily part of the project's requirements. These are nice-to-have items but not a necessity when deciding upon a final solution. Might have features are often characterized as bells and whistles that are part of gold-plating efforts.

6.3 Developing Hybrid Alternatives

After looking at the variables and evaluating all of the alternatives, the conclusion may be that none of the alternatives are acceptable. In this case, the project manager may be forced to select the "best of the worst." As an example, consider a utility which must comply with standards imposed by the Environment Protection Agency. In this case, the company is quite unhappy with all of the alternatives. But, by law, the problem must be resolved and one of the alternatives must be selected.

It is possible that after evaluating the alternatives, the best approach might just be a combination of alternatives. This is referred to as hybrid alternatives. Alternative A might be a high risk but a low cost of implementation. Alternative B might be a low risk but a high cost of implementation. By combining alternatives A and B, we may be able to come up with a hybrid alternative with an acceptable cost and risk factor.

6.4 Phantom Alternatives

Project decision-making usually revolves around the selection of one alternative from a listing of preferred alternatives. A phantom alternative is an option that looks like a viable choice at the time when a decision must be made but for some reason may be unavailable. The phantom alternative could be based upon assumptions or constraints about available resources needed, software development, contractor availability, or favorable testing results. In such cases, the decision-makers selection of a phantom alternative may contain a great deal of risk and uncertainty, and result in ineffective problem resolution.

Decision-making with phantom alternatives is difficult. There are two extreme situations. In the first situation, the options are assumed correctly but to be unavailable in the future. In the second situation, the options are assumed incorrectly but to be available. There are also middle-of-the-road situations.

Phantom alternatives are not always recognizable in the formulation of the problem. Sometimes, an alternative is recognized as a phantom, but decision-makers become highly optimist, hope for the best, and neglect to consider the downstream risks. As such, the phantom may become visible later in the project and results in fewer choices for resolution of a problem.

6.5 Tradeoffs

Tradeoffs are decision-making exercises that most frequently result in changing or sacrificing one part of the project to gain in another part of the project. Hybrid alternatives are examples of tradeoffs. Tradeoffs are most often made on the competing constraints of the project. For example, if the problem is to maintain the quality of the deliverables, we may need to provide more funding, allow for more time, or both. We also use tradeoff analysis techniques when we have several seemingly good alternatives and try to get the best features of each alternative condensed into a single alternative. This involves tradeoffs needed to combine alternatives rather than a tradeoff between alternatives.

While many of the people that sit in the problem-solving and decision-making sessions are experts in their field and come up with realistic alternatives to solve a problem, they often lack the ability to understand the tradeoffs that are necessary and the impact. For example, the solution to meeting the client's quality requirements might be adding more time into the schedule for some additional work, but the cost could be prohibitive. Also, there could be a financial impact on other suppliers that are dependent upon our schedules.

The project team is usually best qualified to evaluate tradeoffs on alternatives even though people outside of the team were used to identify the alternatives. People outside of the project team may be brought in to address a specific problem and may not see the entire picture and the impact of their recommendations.

6.6 Common Mistakes When Developing Alternatives

The alternative selection process is subject to errors as is any other problem-solving process. The errors could be by chance or intentional. Examples of errors that plague the selection process include:

- The time and cost estimates for a particular alternative are grossly underestimated to make this alternative look highly attractive to the decision-makers. The people that provided the estimates may have done this intentionally because of personal gains or recognition if this alternative is selected. This is being done at the expense of the project.

- Support for an alternative is done with overoptimism to the point where the true implementation risks are hidden. This could be the most expensive alternative and the client will then be required to fund lucrative scope changes.
- Support for a good alternative is done with underoptimism in hopes that the client with pick one of the more costly alternatives.

6.7 Decision-Making for Managing Scope Changes on Projects[1]

Not all projects have problems where the solution requires a scope change. For those projects that are contracted projects, contractual policies exist for the process of approving decisions especially if scope changes are required. The type of contract and the terms and conditions may dictate what types of scope changes are allowed.

The number and impact of scope changes on projects vary widely depending on the nature of individual projects and customers' needs. Not all scope changes are the result of problems, but they do require decision-making. Contract changes can present opportunities and challenges that can make or break a project. They can provide additional sales and increased profits. If improperly handled, they can have an adverse effect on profits and on management's reputation. Realization of their potential represents a challenge to the entire organization.

Additional work performed without contract adjustment always results in loss of fee or profit for that additional work. On a cost-plus-fixed-fee (CPFF) contract, this results in reduction of percentage fee on the total contract. On a cost-incentive contract, it results additionally in sharing a percentage of the costs out of target fee. On a fixed-price incentive contract, once the break point has been reached, or on a firm fixed-price contract once the contract value is exceeded, all additional costs come out of company profits. This could result in contractors not wanting to recommend the best solution to a potential problem.

On contracts with the potential for numerous scope changes, a company's overall image and reputation may well be determined by its success in identifying and controlling changes properly. At the same time, considerable maturity of judgment must be exercised to assure that a company's image of competent administration and performance is not defaced by frequent change requests to correct technical deficiencies or contractual oversights. Generally speaking, technical

1 Adapted from Kerzner and Thamhain (1986).

deficiencies and contractual oversights should be brought to the customer's attention, possibly through the medium of "no cost" changes, to avoid possible embarrassment to both customer and contractor.

General Factors Influencing Project Scope Changes

There are numerous factors that determine the number and extent of contract changes on a particular project. Consequently, projects run the gamut from no changes to a flood of changes. Some of the factors that influence the extent of contract changes include:

- **The Customer:** The nature and volume of change problems can be markedly influenced by the particular customer, particular individuals and their attitudes, and practices relative to changes. This occurs frequently on projects where the customer is not sure of the effort needed to satisfy their needs or when an individual has a hidden agenda that they wish to fulfill. Every effort should be made to discern these factors early in the contract and to be guided accordingly.
- **Life Cycle Phases:** The nature, cost, and schedule impacts of contract changes vary considerably as the project progresses through its "life cycle" phases from concept formation to full-scale execution. For any given change, the earlier the change is made, the less will be its cost and schedule impact. A repeat order of items to be manufactured would normally be expected to experience a minimum of change.
- **Concurrency of Life Cycle Phases:** The term concurrency has come into increased usage by some customers to describe a contract situation that requires life cycle phases to overlap. This can significantly increase project risks and create additional problems if work that is best to be performed in series is requested to be performed in parallel. For example, hardware manufacturing that is begun prior to completion of testing and prototype studies can be expected to increase the volume of problems and resulting scope changes.
- **Multilocations and Subcontractors:** The involvement of more than one operating location, major subcontractors, and associate contractors will result in multiple interfaces which may create changes internally and possibly at the customer interface. As an example, based upon the interaction of subcontractors, a consensus might not occur for a scope change that impacts each party differently.
- **Project/Program Size and Duration:** The magnitude of changes will normally increase with the size of the project/program.
- **Type of Contract:** The type of contract can influence the extent of change. Considering a fixed-price contract in contrast to a CPFF contract, it is likely that the parties involved on both sides will be more sensitive to scope definition and the impact of scope changes.

- **Timing of Change:** In relation to many changes, there is a critical time at which they should be officially approved and initiated. In such cases, project issues and costs may increase rapidly if approval and initiation of work are delayed beyond this critical point in time.

Timing for Problem Identification

Early recognition of potential problems that could result in a scope change is important. There is a tendency in the early stages of a project to minimize the importance of minor changes and to consider that they will not have a serious impact on the project. Coupled with this attitude, there is a tendency toward easy acceptance of the view that a change or modification of requirements probably was or should have been considered a part of the initial requirements. Every attempt should be made to discourage such views from the outset and to emphasize the potential cumulative value of even small individual changes and the necessity for establishing proper practices and procedures for identifying and processing problems that could lead to larger changes as the project progresses. At the opposite extreme, however, spending money and irritating the customer over real trivia must be avoided. The project manager is the individual responsible to decide what changes should be pursued on a contract and must establish procedures so that prospective changes are brought to his/her attention for consideration.

Another tendency which prevails in the early stages of the project, and which may be commonly encouraged by the customer is an easy acceptance of an open-ended philosophy with respect to the scope of work. Some customers will act as if the original contract envisioned a progression of improvements to be undertaken at no extra cost. We tend toward easy acceptance of the view that "we should have known or understood that a particular item was required," without serious examination of what the contract says. Every effort should be made from the outset to establish contract baselines that clearly define and limit what is required and to ensure that company personnel understand and restrict their efforts to within those limits.

A source of difficulty, particularly during the systems engineering or development engineering phases, changes in or additions to scope of work which may go on undetected. These changes may come about in a variety of ways. They may result from customer technical direction or suggestion without the proper contractual support. They may come about due to internal interface problems between the company and its subcontractors. They quite commonly occur simply due to the natural and inherent tendency to continue work upon and to make improvements in design. Without special effort, such changes do not necessarily create problems that need resolution but are difficult to detect and may merit special review and investigation by an individual or group assigned this responsibility.

On a large engineering project, the potential exists for identifying situations that can lead to scope changes. Consideration should be given to assigning an individual or group responsibility to review actual work in process for possible change in scope. All technical meetings with customer personnel should be documented and responsibility assigned for review of such documentation for changes in scope. Meeting minutes should also be required for technical meetings crossing major company interfaces as well as interfaces with other contractors or subcontractors.

Projects that have the propensity for many scope changes usually consider special organizational assignments as part of the team. Depending upon the team's actual experience and the potential scope of the changes involved, it may be necessary to assign one or more individuals or groups to handle a variety of functions associated with changes. If the new business potential is sufficient, a specific individual or group may be indicated to study additional or expanded program requirements leading to proposals to be submitted to the customer.

When there is doubt as to whether a scope change has in fact occurred, there might arise a question of whether or not a claim should be pressed. In many cases, it will be quite clear, while in others, the validity of the claim will be questionable. Questionable claims should be resolved at an upper management level based upon a full assessment of the probability of success, the dollars involved, and the impact on customer relations. Borderline cases should not be dropped so long as the possible benefits to the company appear to outweigh the potential disadvantages of pressing the claim.

Discussion Questions

1. What would you expect to be a reasonable number of alternatives for solution to a problem?
2. What are some of the common project management factors used to evaluate alternatives?
3. What are the three most common boundary conditions when discussing the importance of features contained in an alternative?
4. What is an example of a hybrid alternative?
5. What is an example of a phantom alternative?

Reference

Kerzner, H. and Thamhain, H.J. (1986). *Project Management Operating Guidelines, Directives, Procedures, and Forms*. New York: Van Nostrand Reinhold. Chapter 11.

7

Problem-Solving Creativity and Innovation

7.0 The Need for Problem-Solving Creativity

Not all project problems can be resolved by simply changing the resources or modifying the budget and schedule. Some projects have issues that may require significant levels of creativity for a satisfactory solution.

Project managers generally negotiate for resources that can perform the required tasks, assuming that everything will go as planned. The staffing negotiations process does not generally consider whether the resources have the necessary creative skills to resolve challenges that may occur.

As project management expands to include new types of projects, such as those requiring strategic and innovative outcomes, the staffing process must consider what might go wrong on the project and how issues requiring creativity will be managed.

7.1 Creativity and Creative Thinking

You are placed in charge of a project that is quite complex and perhaps even a high risk. Some sort of technical breakthrough may be required for the project to be regarded as a success. Your team keeps coming up with problems that need to be resolved. How does a project manager know if the assigned resources have creative abilities? This is an essential skill for several types of problem-solving.

Not all people are creative even if they are at the top of their pay grade. People can do the same repetitive task for so long that they are considered as subject matter experts. They can rise to the top of their pay grade based upon experience and years of service. But that alone does not mean that they have creative skills. Most

Project Based Problem Solving and Decision Making: A Guide for Project Managers,
First Edition. Harold Kerzner.

Companion Website: www.wiley.com/go/kerzner/projectbasedproblemsolving

people think that they are creative when, in fact, they are not. Companies also do not often provide their workers training in creative thinking.

In a project environment, creativity is the ability to use one's imagination to come up with new and original ideas or things to meet requirements and/or solve problems. People are assigned to project teams based upon experience. It is impossible for the project manager, and sometimes even the functional managers, to know whether these people have the creative skills needed to solve problems that can arise during a project. Unless you have worked with these people previously, it is difficult to know if people have imagination, inspiration, ingenuity, inventiveness, vision, and resourcefulness, all being common characteristics of creativity.

7.2 Creativity and Innovation Thinking

In project management, creativity is the ability to think up ideas to produce something new through imaginative skills, whether a new solution to a problem, or a new method or device. Innovation is the ability to solve the problem by converting an idea into reality, whether it is a product, service, or any form of deliverable for the client. Innovation goes beyond creative thinking.

Creativity and innovation do not necessarily go hand-in-hand. Any problem-solving team can come up with creative solutions that cannot be implemented. Any engineering team can design a product (or a modification to a product) that manufacturing cannot build.

Although most people seem to believe that innovation is directly related to discoveries made by the R&D group, innovation also involves contributions made by all business functions of an organization (sales and marketing, finance, operations, etc.) toward the solution to a problem. Simply stated, innovation, as part of problem-solving, is a team effort.

7.3 Creativity, Innovation, and Value

Innovation is more than simply turning an idea into reality. It is a process that creates value. Clients are paying for something of value. Whatever solution is arrived at must be recognized by the client as possessing value. The best of all possibilities is when the real value can be somehow shared between the client's needs and your company's strategy. The final alternative selected might increase or decrease the value of the end deliverables, as seen by the client, but there must always be some value in the solution selected.

Some solutions to a problem may necessitate a reduction in value compared to the original requirements of the project. This is referred to as negative innovation.

In such cases, innovation for a solution that reduces value can have a negative or destructive effect upon the team. People could see negative innovation as damage to their reputation and career.

If the innovation risks are too great, the project team may recommend some form of open innovation. Open innovation is a partnership with those outside your company by sharing the risks and rewards of the outcome. Many companies have creative ideas for solving problems but lack the innovative talent to implement a solution. Partnerships and joint ventures may be the final solution.

7.4 Negative Innovation

Sometimes, we start out projects with the best of intentions and later discover that some problem has occurred that could result in the cancellation of the project. Rather than cancel the project outright, the solution might be to downsize the project and readjust our innovation attempts. Factors that can lead to a readjustment in innovation include:

- The market for the deliverable has shrunk
- The deliverable will be overpriced, and demand will not be there
- The technical breakthrough cannot be achieved in a timely manner
- There is a loss of faith and enthusiasm by the team, and they no longer believe this solution is workable
- Possible loss of interest by top management and the client
- Insurmountable technical obstacles
- Significant decrease in the likelihood of success.

If these factors exist, then it is entirely possible that another alternative must be selected in order to salvage the project. As long as the client is willing to accept a possible reduction in final value, the project may be allowed to continue.

7.5 Types of Innovation

There are several types of innovation, and the three most common types are listed here. Each type comes with advantages and disadvantages.

- **Product/quality improvements and cost reduction efforts:** this type of innovation may be able to be accomplished quickly and with the existing resources in the company. The intent is to solve a problem and add incremental value to the end result.
- **Radical breakthrough in technology:** this type of innovation has risks. You may not be able to determine when the breakthrough will be made and the

accompanying cost. Even if the breakthrough can be made, there is no guarantee that the client will receive added value from this solution. If the breakthrough cannot be made, the client may still be happy with the partial solution. This type of innovation may require the skills of only one or two people.
- **Totally complex system or platform:** this is the solution with the greatest risk. If the complex system cannot be developed, then the project will probably be considered as a total loss. A large number of highly talented resources are needed for this form of innovation.

7.6 Problem-Solving and Decision-Making Attributes That Are Difficult to Learn

There are courses on problem-solving and decision-making. Unfortunately, there are some attributes that excellent problem-solvers and decision-makers possess, and these cannot be taught easily in courses. The three most common attributes are:

- **Instinct:** the innate behavior that is the inherent inclination toward a particular type of behavior. It is an inborn pattern of behavior that is characteristic of a response to specific stimuli such as the desire to solve complex problems. It is often used with such words as a natural inclination, a natural or acquired tendency, aptitude, talent, knack, gift, capacity, genius, faculty, intuition, feeling, impulse, gut feeling, or sixth sense. People that have this instinct thrive on problem-solving. The more complex the problem, the greater their voluntary involvement.
- **Common sense:** sound, practical judgment derived from experience rather than specialized knowledge. People that possess common sense tend to make decisions based upon the situation and the facts rather than based upon their technical knowledge. However, they may rely upon past experiences.
- **Guesswork:** the process of making guesses, or finding the conclusions arrived at by guessing. Guesswork is often used with other words such as conjecture, deduction, presumption, speculation, estimation, reasoning, approximation, and estimation. Guesswork is required when an estimate must be made and little or no information is available. Based upon the problem, guesswork may be the very first step when initiating problem-solving or decision-making sessions. Sometimes, guesswork is required simply to understand the problem.

7.7 Creativity Roadblocks

There are also blockages to being creative. These include:

- Not understanding the problem well enough and attacking the wrong problem
- Making assessments and decisions too quickly

- Taking the first idea that is acceptable to the team
- Having a team that considers you as an outsider
- Having a team that refuses to support any of your ideas
- Having a team that has no faith in your ability to be part of the team.

These roadblocks do not necessarily apply only to problem-solving that involves innovation. Also, the solution to some problems simply does not require innovation.

Discussion Questions

1. What are examples of project problems that would require creativity for a solution?
2. Can a project manager determine in advance if attendees to a meeting have creativity skills?
3. Can innovative solutions to a problem come from attendees that are in positions that accomplish only repetitive or mundane tasks?
4. What are the differences between creativity and innovation?
5. What is negative innovation?
6. Who is responsible for determining the risks that might accompany a solution requiring innovation?
7. Other than technology, what are examples of creativity roadblocks?

8

Problem-Solving Tools and Techniques

8.0 Root Cause Analysis

Most textbooks do not differentiate between the tools needed for problem-solving and those needed for decision-making because most of the tools can be used for both. Root cause analysis (RCA) and brainstorming are most frequently used for problem-solving and will be discussed here. The other tools, many of which apply to problem-solving, will be discussed later under Chapter 9, Decision-Making.

RCA is a class of problem-solving methods aimed at identifying the root causes of problems or events. The practice of RCA is predicated on the belief that problems are best solved by attempting to address, correct, or eliminate root causes, as opposed to merely addressing the immediately obvious symptoms. By directing corrective measures at root causes, it is more probable that problem recurrence will be prevented. However, it is recognized that complete prevention of recurrence by one corrective action is not always possible. Conversely, there may be several effective measures (methods) that address the root cause of a problem. Thus, RCA is often considered to be an iterative process and is frequently viewed as a tool of continuous improvement.

RCA is typically used as a reactive method of identifying event(s) causes, revealing problems, and solving them. Analysis is done *after* an event has occurred and the problem is visible. Insights in RCA may make it useful as a proactive method. In that event, RCA can be used to *forecast* or predict probable events even *before* they occur. While one follows the other, RCA is a completely separate process.

RCA is not a single, sharply defined methodology; there are many different tools, processes, and philosophies for performing RCA analysis. However, several very broadly defined approaches or "schools" can be identified by their basic

Project Based Problem Solving and Decision Making: A Guide for Project Managers,
First Edition. Harold Kerzner.

Companion Website: www.wiley.com/go/kerzner/projectbasedproblemsolving

approach or field of origin: safety-based, production-based, process-based, failure-based, and systems-based.

8.1 General Principles of Root Cause Analysis

There are general principles that describe RCA. They include:

- The primary aim of RCA is to identify the root cause(s) of a problem in order to create effective corrective actions that will prevent that problem from ever recurring, otherwise addressing the problem with virtual certainty of success. ("Success" is defined as the near-certain prevention of recurrence.)
- To be effective, RCA must be performed systematically, usually as part of an investigation, with conclusions and root causes identified and backed up by documented evidence. Usually, a project team effort is required.
- Although there may be more than one root cause for an event or a problem, the difficult part is demonstrating the persistence and sustaining the effort required to develop them.
- The purpose of identifying all solutions to a problem is to prevent recurrence at the lowest cost in the simplest way. If there are alternatives that are equally effective, then the simplest or lowest-cost approach is preferred.
- Root causes identified depend on the way in which the problem or event is defined. Effective problem statements and event descriptions (as failures, for example) are helpful, or even required.
- To be effective, the analysis should establish a sequence of events or timeline to understand the relationships between contributory (causal) factors, root cause(s), and the defined problem or event to prevent it in the future.
- RCA can help to transform a reactive culture (that reacts to problems) into a forward-looking culture that solves problems before they occur or escalate. More importantly, it reduces the frequency of problems occurring over time within the environment where the RCA process is used.
- RCA is a threat to many cultures and environments. Threats to cultures often meet with resistance. There may be other forms of management support required to achieve RCA effectiveness and success. For example, a "non-punitory" policy toward problem identifiers may be required.

8.2 Corrective Actions Using Root Cause Analysis

RCA forms the most critical part of successful corrective action because it directs the corrective action at the true root cause of the problem. The root cause is secondary to the goal of prevention, but without knowing the root cause, we cannot determine what an effective corrective action for the defined problem will be. Steps to consider include:

- Define the problem or describe the event factually
- Gather data and evidence and then classify that along a timeline of events to the final failure or crisis.
- Ask "why" and identify the causes associated with each step in the sequence toward the defined problem or event.
- Classify causes into causal factors that relate to an event in the sequence, and root causes, that if applied, can be agreed to have interrupted that step of the sequence chain.
- If there are multiple root causes, which is often the case, reveal those clearly for later optimum selection.
- Identify corrective action(s) that will with certainty prevent recurrence of the problem or event.
- Identify solutions that are effective, prevent recurrence with reasonable certainty with consensus agreement of the group, are within your control, meet your goals and objectives and do not cause or introduce other new, unforeseen problems.
- Implement the recommended root cause correction(s).
- Ensure effectiveness by observing the implemented recommendation solutions.
- Other methodologies for problem-solving and problem avoidance may be useful.

8.3 Root Cause Analysis Techniques

RCA tools and techniques include:

- **Barrier analysis:** a technique often used in process industries. It is based on tracing energy flows, with a focus on barriers to those flows, to identify how and why the barriers did not prevent the energy flows from causing harm
- Bayesian inference
- **Causal factor tree analysis:** a technique based on displaying causal factors in a tree structure such that cause-effect dependencies are clearly identified
- **Change analysis:** an investigation technique often used for problems or accidents. It is based on comparing a situation that does not exhibit the problem to one that does, in order to identify the changes or differences that might explain why the problem occurred
- **Current Reality Tree (CRT):** a method developed by Eliahu M. Goldratt in his theory of constraints that guides an investigator to identify and relate all root causes using a cause-effect tree whose elements are bound by rules of logic (Categories of Legitimate Reservation). The CRT begins with a brief list of the undesirable things we see around us and then guides us toward one or more

root causes. This method is particularly powerful when the system is complex, there is no obvious link between the observed undesirable things, and a deep understanding of the root cause(s) is desired
- Failure mode and effects analysis
- Fault tree analysis
- **5 Whys:** ask why, why, why, why, and why over again until exhausted
- Ishikawa diagram, also known as the fishbone diagram or cause-and-effect diagram. The Ishikawa diagram is used by project managers for conducting RCA as well as for resolving quality and risk problems
- Pareto analysis "80/20 rule."

Other tools and techniques exist in addition to what is listed here.

8.4 Brainstorming

Throughout the life of any project, the team will be tested on their ability to find the best possible solution to a problem within the imposed limitations and boundaries. This could occur in the planning phase of the project where we must come up with the best possible approach for a plan or it could happen in any later phases where problems arise, and the best solution must be found. These are situations where brainstorming techniques may be appropriate. Most people seem to have heard about brainstorming but very few have been part of brainstorming teams.

Although brainstorming has become a popular group technique, when applied in a traditional group setting, researchers have not found evidence of its effectiveness for enhancing either the quantity or quality of ideas generated. Although traditional brainstorming does not increase the productivity of groups (as measured by the number of ideas generated), it may still provide benefits, such as boosting morale, enhancing work enjoyment, and improving teamwork. Thus, numerous attempts have been made to improve brainstorming or use more effective variations of the basic technique.

Although we normally discuss brainstorming as a means for identifying alternative solutions to a problem, brainstorming can also be used for root cause identification of the problem.

8.5 Rules for Brainstorming

There are four basic rules in brainstorming. These rules are intended to stimulate idea generation and increase overall creativity of the group while minimizing the inhibitions people may have about working in groups.

- **Focus on quantity:** This rule focuses on the maximization of possible ideas, both good and bad. The assumption made is that the greater the number of ideas, the greater the chance of finding the optimal solution to a problem.
- **Withhold criticism:** In brainstorming, criticism of ideas creates conflict and wastes valuable time needed to generate the maximum number of ideas. When people see ideas being criticized, they tend to withhold their own ideas to avoid being criticized. Criticism should take place but after the brainstorming session is completed. Typical brainstorming sessions last about an hour or less.
- **Welcome unusual ideas:** All ideas should be encouraged, whether good or bad. People must be encouraged to think "out of the box" and this may generate new perspectives and a new way of thinking. Sometimes, what appears as a radical solution initially may be the best possible solution in the end.
- **Combine and improve ideas:** The best possible solution may be a combination of ideas. New ideas should be encouraged from the combination of ideas already presented.

8.6 Critical Steps in Brainstorming

There are several critical steps that must occur for brainstorming to be successful. Simply putting a group of people in a room and saying, "Let's come up with great ideas", does not work well. Professional facilitation and session structure are necessary to maximize performance expectations. Conducting a brainstorming session is not the same as holding a weekly or monthly team meeting. Some of the things that a company can do include:

- First and foremost, it is best to have a professionally trained facilitator conduct the brainstorming session to get people to contribute ideas, bring order from chaos, and limit distractions.
- Send out an agenda early clearly stating the purpose of the meeting, the ground rules, and the topic(s) to be discussed.
- If handouts will be used in the session, it may be best to provide the handouts with the agenda so people can review them prior to the meeting and then come prepared to ask the right questions and possibly make decisions.
- Clearly articulate the reason for the meeting and the goal and make sure that the goal is reachable in a reasonable time frame.
- While asking people to "think outside of the box" seems like a good idea, the best solution may be when the participants think "inside the box" instead.
- Invite participants that may have an interest in the topic even though they are not part of the project team.

- Ask people not to bring distractions such as cell phones, notepads, or laptops.
- Do not criticize any ideas no matter how bad they sound.
- If market research is required, ask the participants to obtain information from the end users rather than the middle people.
- Encourage everyone to come prepared to speak and to share their ideas, whether good or bad.
- Document all ideas because some ideas may be valuable later for issues on other projects. There are several excellent software packages that deal with idea management and brainstorming activities.
- Some people are combative and continuously fight for their belief or their dislike for someone else's position. These people must be controlled to prevent the meeting from losing the intended purpose.

8.7 Conducting the Brainstorming Session: The Process

The process of conducting a brainstorming session includes the following:

- Participants who have ideas but were unable to present them are encouraged to write down the ideas and present them later.
- The idea collector should number the ideas, so that the chairperson can use the number to encourage an idea generation goal, for example: *We have 14 ideas now, let's get it to 20!*
- The idea collector should repeat the idea in the words he or she has written verbatim to confirm that it expresses the meaning intended by the originator.
- When many participants are having ideas, the one with the most associated idea should have priority. This is to encourage elaboration on previous ideas.
- During a brainstorming session, functional managers and other superiors may be discouraged from attending, since it may inhibit and reduce the effect of the four basic rules, especially the generation of unusual ideas.

8.8 Conducting the Brainstorming Session: Evaluation

Brainstorming is not just about generating ideas for others to evaluate and select. Usually, the group itself will, in its final stage, evaluate the ideas and select one as the solution to the problem proposed to the group.

- The solution should not require resources or skills the members of the group do not have or cannot acquire.
- If acquiring additional resources or skills is necessary, that needs to be the first part of the solution.

- There must be a way to measure progress and success.
- The steps to carry out the solution must be clear to all, and amenable to being assigned to the members so that each will have an important role.
- There must be a common decision-making process to enable a coordinated effort to proceed and to reassign tasks as the project unfolds.
- There should be evaluations at milestones to decide whether the group is on track toward a final solution.
- There should be incentives for participation so that participants maintain their efforts.

8.9 Brainstorming Sessions: Nominal Group Technique

There are several variations in the way brainstorming sessions are conducted. A common method is the nominal group technique.

- The nominal group technique is a type of brainstorming that encourages all participants to have an equal say in the process. It is also used to generate a ranked list of ideas.
- Participants are asked to write their ideas anonymously. Then the moderator collects the ideas and each is voted on by the group. The vote can be as simple as a show of hands in favor of a given idea. This process is called distillation.
- After distillation, the top-ranked ideas may be sent back to the group or to subgroups for further brainstorming. For example, one group may work on the color required in a product. Another group may work on the size, and so forth. Each group will come back to the whole group for ranking the listed ideas. Sometimes ideas that were previously dropped may be brought forward again once the group has reevaluated the ideas.
- It is important that the facilitator be trained in this process before attempting to facilitate this technique. The group should be primed and encouraged to embrace the process. Like all team efforts, it may take a few practice sessions to train the team in the method before tackling the important ideas.

8.10 Brainstorming Sessions: Group Passing Technique

The group passing technique includes the following:

- Each person in a circular group writes down one idea and then passes the piece of paper to the next person in a clockwise direction, who adds some thoughts. This continues until everybody gets his or her original piece of paper back. By this time, it is likely that the group will have extensively elaborated on each idea.

- The group may also create an "Idea Book" and post a distribution list or routing slip to the front of the book. On the first page is a description of the problem. The first person to receive the book lists his or her ideas and then routes the book to the next person on the distribution list. The second person can log new ideas or add to the ideas of the previous person. This continues until the distribution list is exhausted. A follow-up "read out" meeting is then held to discuss the ideas logged in the book. This technique takes longer, but it allows individuals time to think deeply about the problem.

8.11 Brainstorming Sessions: Team Idea Mapping Method

The team idea mapping method includes the following:

- This method of brainstorming works by the method of association. It may improve collaboration and increase the quantity of ideas and is designed so that all attendees participate, and no ideas are rejected.
- The process begins with a well-defined topic. Each participant brainstorms individually, then all the ideas are merged onto one large idea map. During this consolidation phase, participants may discover a common understanding of the issues as they share the meanings behind their ideas. During this sharing, new ideas may arise by the association, and they are added to the map as well. Once all the ideas are captured, the group can prioritize and/or take action.

8.12 Brainstorming Sessions: Electronic Brainstorming

Electronic brainstorming includes the following:

- Electronic brainstorming is a computerized version of the manual brainstorming technique. It is typically supported by an electronic meeting system (EMS) but simpler forms can also be done via email and may be browser-based or use peer-to-peer software.
- With an EMS, participants share a list of ideas over the Internet. Ideas are entered independently. Contributions become immediately visible to all and are typically anonymous to encourage openness and reduce personal prejudice.

Modern EMS also support asynchronous brainstorming sessions over extended periods of time as well as typical follow-up activities in the creative problem-solving process such as categorization of ideas, elimination of duplicates, assessment and discussion of prioritized or controversial ideas.

- Electronic brainstorming eliminates many of the problems of standard brainstorming, production blocking and evaluation apprehension. An additional advantage of this method is that all ideas can be archived electronically in their original form, and then retrieved later for further thought and discussion. Electronic brainstorming also enables much larger groups to brainstorm on a topic than would normally be productive in a traditional brainstorming session.
- Some web-based brainstorming techniques allow contributors to post their comments anonymously using avatars. This technique also allows users to log on over an extended time, typically one or two weeks, to allow participants some "soak time" before posting their ideas and feedback. This technique has been used particularly in the field of new product development but can be applied in any number of areas where collecting and evaluating ideas would be useful.

8.13 Brainstorming Sessions: Directed Brainstorming

Directed brainstorming includes the following:

- Directed brainstorming, which is similar to a technique called brain writing, is a variation of electronic brainstorming (described previously). It can be done manually or with computers. Directed brainstorming works when the solution space (that is, the criteria for evaluating a good idea) is known prior to the session. If known, that criteria can be used to intentionally constrain the ideation process.
- In directed brainstorming, each participant is given one sheet of paper (or electronic form) and told the brainstorming question. They are asked to produce one response and stop, and then all of the papers (or forms) are randomly swapped among the participants. The participants are asked to look at the idea they received and to create a new idea that improves on that idea based on the initial criteria. The forms are then swapped again, and respondents are asked to improve upon the ideas, and the process is repeated for three or more rounds.
- In the laboratory, directed brainstorming has been found to almost triple the productivity of groups over electronic brainstorming.

8.14 Brainstorming Sessions: Individual Brainstorming

Individual brainstorming includes the following:

- "Individual Brainstorming" is the use of brainstorming on a solitary basis. It typically includes such techniques as free writing, free speaking, word association, and drawing a mind map, which is a visual note taking technique in which people diagram their thoughts. Individual brainstorming is a useful method in creative writing and has been shown to be superior to traditional group brainstorming under many circumstances.

8.15 Question Brainstorming

Question Brainstorming includes the following:

- This process involves brainstorming the *questions*, rather than trying to come up with immediate answers and short-term solutions. This technique stimulates creativity and promotes everyone's participation because no one must come up with answers. The answers to the questions provide the framework for constructing future action plans. Once the list of questions is set, it may be necessary to prioritize them to reach the best solution in an orderly way. Another one of the problems for brainstorming can be to find the best evaluation methods for a problem.

8.16 Reasons for Brainstorming Failure

Most brainstorming sessions simply do not provide the expected results despite starting out with admirable intentions. This is a fact regardless if you are using the waterfall approach, agile, or Scrum. It is not because the brainstorming process does not work, but because the sessions are conducted poorly. Understanding the reasons for brainstorming failure often serve as a motivational force for corrective action. Some of the most critical reasons for failure, especially with on-site sessions, include:

- **Lack of training for the facilitators and attendees:** Most project management training programs discuss brainstorming but never fully train people on the right way for the sessions to be conducted. Project managers can do more harm than good by facilitating brainstorming sessions without being adequately trained. There are people professionally trained in brainstorming practices.

In such cases, the professional should conduct the session and the project manager may simply be a participant, taking notes if necessary, and answering questions. Ideally, everyone should be trained in brainstorming techniques, so they have a good understanding of expectations when attending such sessions.

- **People spend too much time on solutions:** People tend to focus quickly upon solutions without fully understanding the problem, goal, or question presented. While having people come prepared with ideas and solutions seems a good idea, the focus must be on the right question or problem before generating ideas. Sufficient time must be allowed for people to understand why the meeting was called. Even if this is explained in the invitation email, it should be reinforced at the onset of the meeting for alignment to the issues at hand. Solving the wrong problem is a waste of precious time and money.
- **Poorly trained facilitators begin the meeting by immediately asking for ideas:** The meeting should begin with an understanding of the ground rules (such as no distractions or interruptions), creating the right mindset, explaining the expectations on behavior of the participants (following directions), how the meeting will be conducted, and a clarification of the purpose of the meeting. Even though people may have attended brainstorming training sessions previously, taking a few minutes to explain the ground rules for the session is helpful.
- **Failing to consider the fears and apprehensions of the participants:** Some people have an inherent fear of brainstorming sessions and this includes experienced personnel. The apprehensions might include fear of being criticized, fear of being drawn into a conflict, and fear of change if the implementation of some of the ideas might remove one from their comfort zone.
- **Too long of a meeting:** Brainstorming sessions in some companies may be as short as 15 minutes. Meetings that go beyond one to two hours make people edgy and look at the clock hoping for adjournment. A study conducted at the University of Amsterdam showed that, when people work alone, they tend to come up with more ideas than when working in a group. It may be best to ask people to work first alone, or in small groups, to come up with their best ideas and then share the information in larger groups for evaluation.
- **Large groups can stifle creativity:** There is a valid argument that small groups of 5–10 people yield better results than large groups. Jeff Bezos, Amazon's CEO, calls it the two-pizza rule: If the group can eat more than two pizzas, it is too big. The sessions may be composed of subject matter experts and may not include employees that might later be assigned to the implementation of the desired solution. Having a large group does not mean that more ideas will be forthcoming. Some people may feel intimidated by the size of the group and contribute only a limited number of ideas if at all. With large groups, people worry about how others will view their ideas and may be afraid to contribute for fear of criticism. People tend to contribute more meaningful results when in

small groups. If the group is large, it may be best to break the participants into smaller groups at the start of the session for idea generation and then in a large group near the session's end for idea evaluation. Strong leadership is necessary to prevent the loudest voices and largest groups from drowning out the smaller voices and individualism.

- **Not building on the ideas of others:** Sometimes combining ideas may generate the best possible solution. For this to work correctly, people must be given ample time to express their thoughts and digest what they heard. This is the reason why smaller groups at first are often best. The purpose of the larger group might be for ways that more than one idea from the smaller groups can build on one another for an optimal solution.
- **Having the wrong balance of experience and knowledge in the session:** People should be invited to attend based upon the contribution that they can make rather than simply because of their rank, title, or availability. Inviting people that have a valid interest in the topic, even if they are not part of the project team, may bring forth good ideas.
- **Not having a diverse group:** Having a diverse brainstorming team may be advantageous. More information may come forth that leads to a different and better solution to a problem. Diverse teams usually do a better job challenging the assumptions, looking at problems and solutions differently, and dividing up the work requirements. Each member of the diverse group may come up with good ideas for parts of the solution and, when all the pieces are assembled, a good solution may result.
- **Allowing one person to dominate the discussion:** Some people like to hear themselves talk and try to dominate the discussion. This can be demoralizing to others and the frustration can prevent others from wishing to speak.
- **Information overload:** Brainstorming sessions run the risk of information overload. This is particularly true if multiple brainstorming sessions are needed. Information overload can be demoralizing but can be controlled using idea management software.
- **Premature evaluation:** Some groups tend to quickly jump on the first acceptable idea and run with it without proper evaluation. Forcing participants to vote without due consideration of the facts can result in the implementation of a suboptimal solution that everyone will question.

8.17 Virtual Brainstorming Sessions

The previous discussion assumed that the people attending the brainstorming session were in the same building or location. Today, workers generally spend a minimum of 25% of their time in virtual communications. This percentage is increasing as companies are now embarking upon virtual brainstorming sessions because more people are working from home because of several factors including COVID-19, the cost of office rental space, and the workers needed for the sessions may be dispersed geographically across multiple continents.

Virtual brainstorming is somewhat more difficult than on-site brainstorming because the virtual environment may require a different set of tools and software for communication, viewing, recording and displaying of ideas, and interaction among participants. If the group must be broken down in smaller groups, multiple concurrent virtual sessions may be necessary. The proper use of virtual brainstorming tools can overcome the productivity loss encountered in on-site brainstorming, bring out more creative ideas per person, and generate a higher degree of satisfaction among the team members.

Virtual brainstorming has advantages and disadvantages, many of which are like on-site brainstorming. The benefits include:

- The workers are under less peer pressure and may not be intimidated by others on the call.
- It may be easier to put together a diverse team of participants.
- People are working alone or in small groups and may come up with more fruitful ideas than in larger groups.
- Large groups can participate virtually, and it is less likely that someone will want to dominate the discussion with their ideas.
- Large groups can be subdivided into smaller groups without worrying about title, rank, and expertise.
- There is less wasted time in virtual sessions than on-site sessions.

There are several disadvantages including:

- Facilitators must ensure that the proper virtual tools are in place.
- It may take more time at the onset of the meeting to make sure that everyone is on the same page.
- Sharing documents may be difficult virtually; facilitators must ensure that all participants have the appropriate handouts.
- The way that communication takes place may make it difficult for workers to build on the ideas of others or to combine ideas.
- It may be difficult to break large groups into smaller groups virtually.
- Virtual participants may be less likely to ask questions than if they were in the conference room with the other team members.
- Professional facilitators are trained in emotional intelligence and how to read body language. They can observe the expressions on people's faces and watch what they do with their hands or the way they are sitting as an indication of whether they are upset or in agreement with the discussion. Their fears and apprehensions can be visible by how they act. This is difficult to observe virtually.

- Having an open dialogue where everyone gets to speak may be difficult to enforce.
- Having too large a group may prevent or discourage members from providing input.
- People may be multitasking or distracted, and the facilitator has limited control over the meeting.

Discussion Questions

1. What are the advantages and disadvantages of conducting a RCA?
2. Why do many brainstorming sessions fail to achieve optimal results?
3. What are typical rules for conducting a brainstorming session?
4. Should agendas and handouts be provided in advance to brainstorming session attendees?
5. How many attendees should there be in brainstorming sessions and who makes the determination?
6. Should time limits be set for a brainstorming session, and if so, what are typical limits?
7. Why are there so many types of brainstorming sessions?

9

Decision-Making Concepts

9.0 Decision-Making Alternatives

Today, project teams are faced with difficult-to-solve problems that are based upon increasing levels of complexity. Years ago, project managers were expected to use judgment, bargaining, or analytical approaches to solve problems. Judgment was performed by decision-makers that trusted their own intuition and did not feel the need of justifying their reasoning to the team or even governance personnel. Bargaining was used when project managers were unsure of the best approach and used a team consensus to reach a decision. The team consensus also included input and influence from powerful stakeholders. Analytics were used if time permitted and if the organization had decision-making software available. Analytics were often avoided on projects that were highly complex and with a great deal of risk and uncertainty.

Over the years, sponsors and governance committees provided project teams with a great deal of authority to make their own decisions without senior management's involvement. As the paperwork and information requirements for decision-making increased, senior management was reluctant to spend a great deal of time reading through the documentation and evaluating the information for most project decisions as it was taking time away from their other responsibilities.

Project teams today have several approaches available to them for problem solving and decision-making. Some of the new techniques come from management science and operations research models. Other techniques can be more formal models, methods, and procedures that are part of project management methodologies and frameworks.

Project Based Problem Solving and Decision Making: A Guide for Project Managers,
First Edition. Harold Kerzner.

Companion Website: www.wiley.com/go/kerzner/projectbasedproblemsolving

9.1 Decision-Making Characteristics

We must now decide which of the alternatives best resolves the problem. More often than not, decision-making has some degree of structure to it. Decision-making involves the following:

- Objectives must first be established
- Objectives must be classified and placed in order of importance
- Alternative actions must be developed
- The alternatives must be evaluated against all the objectives
- The alternative that can achieve all the objectives is the tentative decision
- The tentative decision is evaluated for more possible consequences
- The decisive actions are taken and additional actions are taken to prevent any adverse consequences from becoming problems and starting both systems (problem analysis and decision-making) all over again.

The decision-making activities are often more time-consuming and costlier to perform than the problem-solving activities. This is largely due to the number of alternatives that can be identified, and the methods used to evaluate and prioritize them. Having a significant number of reasonable alternatives may seem nice but being unable to arrive at a decision on which one to actually adopt can be troublesome.

There are several types of decision-making styles that people can use. There are also numerous tools that can assist in the decision-making processes.

9.2 Decision-Making Participation

Decision-making alternatives selected by the project manager must be based upon the project manager's determination of which people, if any, should participate in making the decision. There are several reasons why the project manager may be inclined to make the decision by himself/herself. Examples might include the need for rapid response, the problem is well-defined and with limited choices for resolution, there are limited resources available that are familiar with this type of problem, the team expects the project manager to make the decision alone and the team will accept the decision, and having the team participate in the decision might lead to conflicts among the team members resulting in elongation of the project.

There are likewise several reasons why the project manager would want participation by others. Examples might include the project manager's lack of knowledge of all the alternatives, expectations, and desires of team members to participate in decision-making, and the expectation that team member participation will serve as a motivational force for successful execution of a solution.

9.3 Understanding How Decisions Are Made

People must understand prior to attending both the problem-solving and the decision-making meetings how the decisions will be made. There are several options available, and the approach taken to agree on the problem can be different from the decision on which alternative will be adopted. Options include:

- **Majority or consensus:** all participants in the meeting are allowed to vote. The criteria might be a simple majority or another number such as a 75% majority.
- **Qualified majority or consensus:** if a majority is not reached, then the project manager, the client, or another designated individual will make the final decision.
- **Project manager directed:** the project manager makes the decision and informs the team which alternative he/she selected. This approach is most effective on crisis projects.
- **Executive directed:** the executives or project sponsors make the decision based upon the success metrics that they believe track the project's success criteria. If the definition of success changes over the course of the project, this could create a problem.
- **Client directed:** the team identifies the alternatives, make a recommendation, and presents the data to the client. The client then makes the final decision and informs the team. The client may have the right not to select from the team's alternatives but to develop their own solution to the problem.

9.4 Cultures and Problem Solving

All companies have project problems needing resolution, but some companies find the problem-solving process easier than others due largely to their culture. Culture is the shared values, beliefs, and performances that define a group of people and their behavior. Organizational cultures are intangible assets that can lead a firm to success or failure. Companies that are highly successful in project management have cooperative cultures which support problem solving and decision-making.

Companies that struggle appear to have project management cultures that are not necessarily aligned to the corporate culture. In these companies, the corporate culture usually focuses heavily upon command and control from the senior levels of management and project team members feel somewhat restricted in expressing their ideas for solving a problem for fear of criticism from above and possibly being reprimanded for coming up with a bad idea.

Cooperative cultures thrive on free thinking and encourage people to think outside of the box and outside of their area of expertise when necessary. Good cooperative cultures strike a balance between order and chaos. Some characteristics of cooperative cultures include:

- Executives encourage project teams to express themselves without fear of retribution or punishment
- Executives share strategic information with project teams to help align solutions to strategic project objectives
- Executives participate when needed in project problem-solving and decision-making sessions without dominating the discussions
- Executives encourage project teams to use whatever decision-making style is most comfortable to the team.

9.5 Routine Decision-Making

Some decisions are easy to make whereas others require teams of experts. The tools and techniques used are dependent upon the type of decision, and there are numerous types of decisions that can be made.

Routine decisions are often handled solely by the project manager. Routine decisions may involve simply signing purchase orders, selecting which vendors to work with, and deciding whether or not to authorize overtime. Usually, routine decisions are based upon company policies and procedures.

While routine decisions seem relatively easy to make, the number of routine decisions can be troublesome. Too many routine decisions can become time robbers and prevent the project manager from effectively managing the project. If the decisions are routine in nature, then many of the decisions may be able to be delegated to members of the project team.

9.6 Adaptive Decision-Making

Adaptive decision-making may require some degree of intuition. The problem is usually well understood, and the project team may be able to make the decision without outside support or sophisticated tools and techniques. Adaptive decision-making is the most common form of decision-making used on projects. Examples might include:

- Determining the number of tests that should appear in the test matrix
- Determining when an activity should begin or end
- Determining how late an activity can start without delaying downstream work

- Determining how late we can order raw materials
- Determining whether the work should take place on regular shift or overtime
- Determining whether a risk management plan is necessary, and if it is necessary, how much detail should appear in the place
- Determining how often testing should take place to validate compliance to quality requirements
- Determining the resource skill set needed, assuming there are choices
- Determining the best way to present both good news and bad news to the stakeholders
- Determining ways to correct unfavorable cost and schedule variances
- Determining the leadership style to be used to motivate certain team members
- Determining how to best reward superior performance by team members.

9.7 Innovative Decision-Making

Innovation is generally regarded as a new way of doing something. The new way of doing something should be substantially different from the way it was done before rather than a small incremental change such as with continuous improvement activities. The ultimate goal of innovation is to create hopefully long-lasting additional value for the company, the users, and the deliverable itself. Innovation can be viewed as the conversion of an idea into cash or a cash equivalent.

Innovative decision-making is most often used on projects involving R&D, new product development, and significant product enhancements. These decisions involve subject matter experts that may not be part of the project team and may require the use of more advanced decision-making tools and techniques. These decisions may require a radical departure from the project's original objectives. Not all project managers are capable of managing projects involving innovation by themselves.

While the goal of successful innovation is to add value, the effect can be negative or even destructive if it results in poor team morale, an unfavorable cultural change, or a radical departure from existing ways of doing work. The failure of an innovation project can lead to demoralizing the organization and causing talented people to be risk-avoiders in the future rather than risk-takers.

9.8 Pressured Decision-Making

Time is a critical constraint on projects, and this can have a serious impact on the time necessary to understand the problem and find a solution. As an example, let us assume that a critical test fails, and the client says that they will be meeting with you the day after the failure to discuss how you will correct the problem. They are expecting alternatives and a recommendation.

Typically, you might need a week or longer to meet with your team and diagnose the situation. However, given the circumstances, you may have to make a decision, right or wrong, based upon the time available. This is high-pressured decision-making. Given sufficient time, we can all analyze or even overanalyze a problem and come up with a list of viable alternatives.

High-pressured decision-making can also be part of adaptive and innovative decision making as well. Being pressured to make a decision can have favorable results if it forces the decision-makers to look at only those attributes that are critical to the problem. But more often than not, high-pressured decision-making leads to suboptimal results.

Given that these situations will happen, you must expect that you will not always have complete or perfect information in order to make a decision. Most decision-making teams must deal with partial information.

9.9 Judgmental Decision-Making[1]

One type of decision-making that is based upon past experiences or knowledge is a judgmental decision-making. The decision-maker remembers what has happened in similar past situations and uses this database to predict the outcome of the alternatives that are involved in the decision. This is a useful approach to decision-making since many situations, problems, and opportunities do tend to recur frequently in organizations. A judgmental decision is faster and more inexpensive to make; however, it does have a drawback. It is based on common sense and memory, and some people do not have common sense or good memory. Then, too, judgmental decisions are not able to deal with those situations which are new and in which there is no experience base on which to select a logical course of action.

9.10 Rational Decision-Making

Rational decision-making is one in which a rational process is reasoned through to make the decision. Certain conditions must exist before we can say that people are approaching decision-making in a rational fashion. In the first place, there

1 The next five sections have been adapted from Cleland and Kerzner (1986).

must be some objectives and goals toward which resources can be committed. Second, the decision-maker must develop a clear understanding of the alternative actions that can be taken to reach the objectives and goals. Third, they must have the information to analyze the alternatives considering the goals or objectives that are being sought. Finally, the decision-maker must have a desire to come to the best possible solution by selecting the alternative which best satisfies the overall effectiveness of the organization.

It is often difficult to find complete rationality in the decision process. In the first place, decisions are made today but impact the future, and the future almost invariably involves risks and uncertainties, some of which will not be known. Second, all the alternatives that must be analyzed to make a decision may never be recognized, simply because there are many ways in which organizational purposes can be achieved. Finally, not all alternatives can be analyzed even with the most modern analytical techniques, simply because there are so many alternative ways of doing things and the decision-maker is always faced with a time frame for the decision.

9.11 Certainty/Uncertainty Decision-Making

Some forms of decision-making can be done under conditions of certainty, although such conditions are rare. Decision-making under certainty occurs when the manager knows exactly which environmental or competitive action will occur. That is, the manager can make perfectly accurate decisions on a routine basis. Decision-making under certainty requires supporting information that is nearly perfect. Unfortunately, such perfect information is difficult to obtain since some of the variables in a decision, such as how people will react to the decision, is never predictable with complete accuracy.

9.12 Controllable/Noncontrollable Decision-Making

Managers operate within environments in which some of the decision variables are controllable and others are noncontrollable. In a controllable environment, the decision-maker can select a strategy and control the execution of that strategy.

In the decision-making context, the one major variable that is controllable is the actual choice of the strategy. However, the implementation of that strategy will be affected by uncontrollable variables within the environment.

There are two variables which influence the selection of alternatives in a strategy, namely (i) the states of nature and (ii) competitive action. We use the term, "states of nature" to designate the possible events that may actually occur. For the business manager, competitive actions are uncontrollable.

States of nature are uncontrollable from the position of the manager. For example, the accuracy of forecasts will depend in a large part upon the state of the overall economy. In today's complex environment, there are many possible states of nature that can occur. It's virtually impossible to list them all and to determine the effects that each might have on an outcome.

9.13 Programmed/Nonprogrammed Decision-Making

Decisions can be classified as programmed or nonprogrammed.[2] Programmed decisions are standing decisions that exist to guide managers and professionals in their daily organizational life, existing organizational mission, objectives, goals, strategies, policies, standards, procedures, methods, and roles. Programmed decisions are guided by:

- Company policies and guidelines, which are guides set down by higher-level executives for other managers and professionals in dealing with recurring anticipated situations. A product quality policy, a management development policy, and an inventory level policy are examples of policies. Discretion is expected in using policies as guidelines in programmed decisions.
- Standards, which are criteria against something which can be compared. Engineering design standards, manufacturing output standards, and product quality standards are examples of criteria that guide decision-making and implementation that are basically unchanged from day to day.
- Procedures, which are sequential steps to carrying out action in the accomplishment of some tasks. Tasks are "procedurized" to reduce complexity and ensure uniformity in their completion. Little discretionary behavior is permitted in using procedures. For example, the procedure for processing an engineering change on a construction project requires certain sequential actions that must be carried out to evaluate the cost, schedule, technical aspects, and contract implications of the change.

Policies, standards, and procedures provide a framework within which decisions can be made. Such a framework reduces the chance of error. It also reduces the time to make a decision. Prudent managers may delegate authority and

2 See Simon (1977).

responsibility to make decisions and allocate resources within the framework of a policy.

The development by a manager of adequate policies, procedures, and standards permits fuller delegation of decision-making authority and reduces the time the manager must actively supervise professionals. Then "management by exception" to the policies, procedures, and standards can be carried out by the manager, conserving that individual's time.

If an organization has a well-defined strategic framework of mission, objectives, goals, and strategies, then key decisions regarding the allocation of resources can be made within it. This strategic framework becomes a baseline to guide key decision-makers in the enterprise. In this sense, the strategic framework provides a basis for a "programmed" decision, within a framework of organizational mission, objectives, goals, and strategies.

Programmed decisions are those which are made in accordance with some pre-established guidelines, such as in a policy, procedure, rule, or methodology. If problems, opportunities, or issues recur, and if the integral elements can be defined, analyzed, and predicted, then we may develop supporting policies, procedures, methodologies, or rules in order that decision-making can be "programmed." For example, the decisions on an inventory level to support a given product would involve fact finding and forecasting. If the analysis of the inventory level is properly performed, then from that a series of routine decision-making guidelines can be developed. Programmed decisions, however, do limit the freedom of the decision-maker.

As an executive moves up the chain of command within an organization, the ability to deal with unprogrammed decisions becomes more important, since more decisions of a nonprogrammed nature are found at the top level of organizations.

Nonprogrammed decisions are unique and nonrecurring. Although the decision may be made within a strategic framework, the decision-maker is faced with new alternatives and choices. The problem or opportunity is novel, inherently unstructured, and usually involves unknown factors and forces. The selection of an enterprise's strategic framework, its mission, objectives, goals, and strategies are examples of nonprogrammed decisions. However, once established, this strategic framework provides the basis for certain programmed decisions to be made.

In nonprogrammed decisions, many different options are open to the manager. Creativity and innovation on the part of the decision-maker are valued. Nonprogrammed decisions are found in every function of management: planning, organizing, motivating, leading, and controlling. The strategic planning function provides the greatest opportunity for nonprogrammed decisions since the decision-maker is attempting to create something that does not currently exist. Intuition plays an important role in nonprogrammed decisions.

Each manager should develop a set of decision-making guidelines that provide general standards for how much time and analysis to use in evaluating and making different types of decisions.

9.14 Decision-Making Meetings

While some decisions that are routine or adaptive in nature can be handled during regular team meetings, in general, problem resolution team meetings should be set up as separate meetings. The attendance at a problem resolution meeting can be quite different from the attendance at regular team meetings. Stakeholders and clients may be required to attend problem resolution team meetings since they are the people most likely affected by the decision. Functional managers and subject matter experts may also be invited to attend. Outside consultants with critical expertise may also participate.

There is a wide variety of decision-making tools and techniques that project teams can use for decision making during these meetings. The selection of the best tool or technique to use can be based upon the complexity of the problem, the risks associated with the decision, the cost to make the decision and the impact if the decision is wrong, to whom the decision is important, the time available to make a decision, the impact on the project's objectives, the number of people on the project team, the relative importance to the customer or stakeholders, and the availability of supporting data.

Generally speaking, more than one meeting may be required. The purpose of the first meeting may be to just understand the problem and gather the facts. The problem-solving team may then require additional time to think through the problem and identify alternatives. It is highly unlikely that a decision will be made at the first team meeting.

9.15 Decision-Making Stages

There are several models available for how project teams make decisions. A typical four phase model might include:

- **Familiarization stage:** this is where the team meets to understand the problem and the decision(s) that must be made.
- **Options identification phase:** this is where the team performs brainstorming and lists possible alternatives for a solution.
- **Option selection phase:** this is where the team decides upon the best option. The team selecting the preferred option may have a different makeup than the team developing the list of options.

- **Justification phase:** this is where the team rationalizes that they made the right decision and possibly evaluates the results.

9.16 Decision-Making Steps

When a project team is faced with a difficult decision, there are several steps one can take to ensure the best possible solution will be found. The decision process begins and ends with the use of information and judgment in removing the uncertainty of future actions. Decisions do not deal with the past; they deal with the present and the future. A decision-maker needs to be creative and skillful and to operate from an experience base in using information in arriving at decisions.

There are several models that can be used. Many of the steps are similar to the steps used in setting up the problem-solving process. Steps that are common to many of these models include:

- Determine the framework of the problem or opportunity that requires a decision
- Send out an agenda identifying the purpose of the decision-making meeting and the expected outcome, including the expected timing for the decision
- Ask the team to come to the meeting with alternative solutions, if possible
- Perform a root cause analysis of the problem to make sure everyone understands why a decision is necessary, and the impact if a decision is not made
- Collect the appropriate facts and assumptions bearing on the decision
- Analyze the facts and assumptions that impact the problem or opportunity
- Develop alternatives that are viable in dealing with the decision
- Analyze the alternatives based on an assessment of the strengths (benefits) and weaknesses (costs) relative to existing organizational purposes (mission, objectives, goals, and strategies)
- Select the alternative that has the best fit with organizational purposes
- Delineation of a plan of action on how the decision will be implemented.

These steps formulate the decision-making process carried out by managers and professionals. These steps are found with varying degrees of formality where managers and professionals make decisions in organizations. Decision-making requires demanding knowledge, experience, skills, and attitudes in assessing the risks and uncertainties surrounding decisions. Decision situations where the conditions of certainty exist are rare. The decision-maker seldom has full knowledge of all the alternatives and their consequences. The search for alternatives continues until the decision-maker finds an alternative that best suits some personally determined minimum acceptable level. The notion of trying to find the "best" alternative is controversial in the pragmatic real world.

A decision-maker searches for alternatives only if he is dissatisfied with the best alternative he has. There is usually no real attempt made to put the alternatives in a rank order based on value. A favored alternative is usually established early in the decision process and survives through to the final formal decision. Early conclusions concerning the selection are often based on subjective or intuitive factors. There is no question that many managers make decisions with an explicit or implicit desire to "get by" and make a decision that is good enough for the situation.

Many of the decisions contain so many complex variables that it is difficult for an individual to examine all of them fully. In real life, managers make logical decisions, but these decisions are limited by inadequate information and by the ability of the decision-maker to utilize that information. Rather than seeking the best or ideal decision, managers frequently settle for a decision that will adequately serve their purposes. In other words, they "satisfice" or accept the first satisfactory decision they uncover rather than maximizing or searching until they find the optimal decision.[3]

Individuals can develop a series of steps for making decisions. First, they need to develop in their own minds and in their modus operandi a rational approach to the decision-making process. Second, having developed a rational approach to the decision-making process, they should proceed methodically and carefully in their collection and analysis of information and in the delineation of alternatives for the decision. Third, they should get people involved who can help them in the analysis of the information for the decision and in the assessments of the probability of the decision being effectively implemented.

9.17 Conflicts in Decision-Making

Many of the decisions we must make have conflicting values. The decision-maker who occupies a managerial position is always involved in trade-offs in decisions. Two fundamental trade-offs exist: (i) the need to preserve the overall effectiveness in the organization; and (ii) the resolution of the inherent conflict found in organizations because people tend to view the organization, and the decision affecting it, from a provincial viewpoint. The manager must take these provincial viewpoints into consideration and make a decision which best preserves the overall effectiveness of the organization.

Considering the opportunity for conflict and the need for trade-offs, many managers procrastinate and allow these considerations to paralyze their decision-making. These are people who would prefer to let things drift with the tide and

3 Simon (1957). See also March and Simon (1958).

not make a decision. However, these are the situations in which deliberately not making a decision is in fact a decision itself. It means that resources will continue to be consumed in the way that they had been consumed in the past. Thus, no decision is in fact a decision.

9.18 Advantages of Group Decision-Making

As stated before, there are situations such as routine decisions where the ultimate decision is made by just the project manager. Groups are not needed in this situation. But more often than not, the problems that appear on projects require group thinking.

There are several advantages to group decision-making. These include:

- Groups provide better decisions than individuals
- Group discussions lead to a better understanding of the problem
- Group discussions lead to a better understanding of the solution
- Groups make better judgments calls on the selection of alternatives
- Groups tend to accept more risks in problem solving than do individuals
- Clients appear less likely to question the decision of the group compared to the decision of an individual
- People are more willing to accept the final decision if they participated in the decision-making process.

9.19 Disadvantages of Group Decision-Making

There are several disadvantages to group decision-making. These include:

- The discussions can be dominated by the personality of one person regardless of whether or not that person is regarded as a subject matter expert
- Groups may accept too much risk knowing that a failure would be blamed equally among all of the members of the group
- There is pressure to accept the decision of the group even though you know that other decisions might be better
- Too much time may be spent arriving at a consensus
- Groups tend to overthink problems and solutions
- It may be impossible to get the proper people released from other duties so they can attend the meeting
- Finding a common meeting time that satisfies all parties may be difficult
- If external people are involved, the costs associated with traveling could become quite large especially if more than one decision-making meeting is needed.

9.20 Rational Versus Intuitive Thinking

Decision-making requires thinking. There are four forms of thinking that we will discuss. Rational thinking, often referred to as analytical thinking, refers to logical or reasoning being involved in the thought processes required for problem-solving and decision-making. It refers to providing reasons or the rationale behind thoughts or ideas. It adds an element of calculation and planning to a stream of thoughts rather than basing them on emotion or personal opinion. It is a kind of objective process of thinking and an analytic approach to any problem. Rational thinking is based on reasons or facts and is hence much more calculating and realistic. All people are capable of thinking rationally, but people will tend to cloud this ability because of emotions, prejudices, and possibly a fear of making a decision.

Rational thinkers believe that problems are easier to solve if they are broken down into well-defined sequential steps. They maximize the use of forms, guidelines, templates, and checklists. Rational or analytical thinking is efficient in the following conditions – sufficient time, relatively static conditions, and a clear differentiation between the observer and the observed. It is best suited for dealing with complexities and works best where there are established criteria for the analysis (e.g. rules of law).

Intuitive thinking has contrasting qualities: it is unfocused, nonlinear, contains "no time," sees many things at once, views the big picture, contains perspective, is heart-centered, oriented in space and time, and tends to the real or concrete. Intuition comes into its own where analytical thinking is inadequate: under time pressure, where conditions are dynamic, and where the differentiation between observer and observed is unclear. Intuitive thinking is a necessity when working on projects. It works best when seeking the "best" option in favor of the "workable," and when the project team is prepared to act on feelings or hunches where explanations are either not required or there is no time for them. Intuition is experience translated by expertise to produce rapid action.

9.21 Divergent Versus Convergent Thinking

Divergent thinking is a thought process or method used to generate creative ideas by exploring many possible solutions by expanding the problem and looking at the big picture. It is often used in conjunction with convergent thinking, which is a narrow focused detailed picture that follows a particular set of logical steps to arrive at one solution, which in some cases is a "correct" solution. Divergent thinking typically occurs in a spontaneous, free-flowing manner, such that many ideas are generated. Many possible solutions are explored in a short amount of time, and unexpected connections are drawn. After the process of divergent thinking has been completed, ideas and information are organized and structured using convergent thinking. Divergent thinking moves away in diverging directions so as to involve a variety of aspects, which sometimes lead to novel ideas and solutions. It is usually associated with creativity. Idea generation techniques such as brainstorming and out-of-the-box thinking are used in which an idea is followed in several directions to lead to one or more new ideas, which in turn lead to still more ideas.

In contrast to divergent thinking, which is creative, open-ended thinking aimed at generating fresh views and novel solutions, convergent thinking aims at bringing together information focused on solving a special problem (especially solving problems that have a single correct solution).

9.22 The Fear of Decision-Making: Mental Roadblocks

Not everyone wants to make decisions or is capable of making them. Some people would prefer to have others make all decisions, especially critical decisions. Reasons for this behavior might include:

- A previous history of making the wrong decisions
- Emotionally afraid of making the wrong decision
- Afraid of the associated risks
- Lack of conviction in one's own beliefs
- Having high levels of anxiety
- Unable to cope with the politics of decision-making
- Unfamiliar with the facts surrounding the problem and not willing to learn
- Unfamiliar with members of the team
- Possessing poor coping skills
- Lack of motivation
- Lack of perspective
- Being brought into the discussion well after the discussion began
- Unable to work under high levels of stress and pressure
- Afraid of working with unions that are involved in the problem
- Afraid of working with certain stakeholders involved in the problem
- Afraid of contributing for fear of being ridiculed
- Afraid of exposing one's inadequacies
- Afraid of damaging your career and/or reputation.

These roadblocks are often categorized into five areas:

- Emotional blockages
- Cultural blockages
- Perceptual blockages
- Intellectual blockages
- Expressive blockages.

9.23 Decision-Making Personal Biases

Biases can creep into our decision-making processes. A partial list might include:

- Believing beforehand that your solution is the only possible solution
- Ignoring evidence that supports a conclusion other than yours
- Neglecting to understand the root cause of the problem
- Refusing to search for supporting data for a decision
- Neglecting to understand how the wrong decision can impact the project
- Being afraid to state your opinion and siding with the person whom you believe will provide the best approach
- Being afraid to make a decision for fear that you may make the wrong decision
- Being fearful of having your ideas criticized
- Unwilling to think differently or out-of-the-box
- Adopting a wishful thinking approach to making a decision
- Adopting a selective perception approach and looking at only the information and alternatives that are in your comfort zone
- Making the decision that others expect you to make even when you strongly believe it may be the wrong decision
- Making a decision that is in your personal interest rather than the best interest of the project
- Spending too much time on small or unimportant things that are in your comfort zone rather than focusing on what is critical.

9.24 The Danger of Hasty Decisions

The project's constraints often place the project manager in a position of wanting to make hasty decisions. Making hasty decisions is sometimes a necessity, but more often than not, the results can be detrimental. Hasty decisions can lead to:

- Additional problems surfacing later in the project
- Rework that leads to cost overruns and schedule slippages

- Excessive overtime
- Customers and stakeholders will lose faith in your ability to manage the project correctly
- Lack of faith in the problem-solving and decision-making process
- Manpower curves with peaks and valleys rather than a smooth manpower curve
- Greater hands-on involvement by the governance committee
- More meetings
- An increase in reporting requirements
- Deliverables that are rejected by the client.

Simply stated, speed in decision-making is a risky business.

9.25 Decision-Making Styles

Not all decisions are easy to make. Sometimes you must make a decision whether you are ready or not, and when you have partial rather than complete information available to you. Also, the decision to do nothing differently may be the best decision under certain circumstances. If the team believes that they can live with the problem at hand, then the team may wait and see if the problem gets worse before making a decision.

Every project manager has their own approach to decision-making and this may vary from project to project. The style selected is based upon the definition of the problem and the type of decision that must be made. Although some approaches work well, there are approaches that often do more harm than good.

Textbooks on decision-making provide several different styles. The five styles most common for project managers are:

- The autocratic decision-maker
- The fearful decision-maker
- The circular decision-maker
- The democratic decision-maker
- The self-serving decision-maker.

9.26 The Autocratic Decision-Maker

The autocratic decision-maker usually trusts nobody on the team and dictates the decision even though the risks are great and very little time was consumed discussing the problem. Team members often are fearful of presenting alternatives and recommendations because they may be ridiculed by the project manager that believes that his/her decision is the only one. Team members may not contribute ideas even when asked.

The autocratic style can work if the project manager is regarded as an expert in the area in which the decision must be made. But in general, project managers today seem to possess more of an understanding of technology than a command of technology. As such, using the autocratic style when you have limited knowledge about the technology of the problem and the solution can lead to a rapid decision but often a decision that is not the optimal choice.

Most of the time, autocratic decision-makers feel better making a decision by themselves without any input from others. They make a decision on the spot based upon a feeling in their gut. This is often a hit-or-miss approach.

9.27 The Fearful Decision-Maker

While the autocratic decision-maker thrives on making the decision, right or wrong, and in a timely manner, the fearful decision-maker is afraid of making the wrong decision. This is often referred to as the "ostrich" approach to making a decision. In this case, the project manager will bury his/her head in the sand and hope that the problem will disappear or that people will forget about the problem. The project manager also hopes that by waiting, a miracle solution will appear by itself such that a decision may not have to be made at all.

Sometimes the fearful decision-maker adopts a procrastination attitude, which is waiting for enough (or at least a minimum amount of) information so that a decision can be made. This does not necessarily mean avoiding a decision. The fearful decision-maker knows that a decision must be made, eventually.

The fearful decision-maker is afraid that making the wrong decision could have a serious impact upon his/her reputation and career. The team may not be invited to provide alternatives and recommendations because that would indicate that a problem exists and that a decision must be made. Information on the problem may even be withheld from senior management, at least temporarily.

The project manager may try to get others to make the decision. The project manager may prefer to have someone act as the moderator of the decision-making group and, if a decision must be made, the project manager will always argue that it is a group decision rather than a personal decision. The project manager will avoid, if possible, taking personal accountability and responsibility for the decision.

As stated previously, time is a constraint on projects, not a luxury. Taking a wait-and-see approach to making a decision can lose precious time where the problem could have been easily resolved. Also, the longer we wait to make a decision, the fewer the options are.

9.28 The Circular Decision-Maker

The circular decision-maker is similar to the fearful decision-maker. The project manager not only wants to make the decision but wants to make the perfect decision. Numerous team meetings are held to discuss the same problem. Each team meeting seems to discuss the problem and possible solutions from a different perspective. The team members are given action items that keep them scurrying about looking for additional information to support the perfect decision.

The circular decision-maker is willing to make a decision but sacrifices a great deal of time looking for the perfect decision that everyone will agree to. The decision-maker is willing to violate the time constraints on a project to accomplish this. The decision-maker may also believe that the problem may disappear if they think about it long enough.

The project manager can adopt the circular decision-making style even if he/she is an expert in the area in which the problem exists. The project manager needs re-enforcement from the team, and possibly superiors, that the best decision was made. In the eyes of the project manager, the decision may be deemed more important than the outcome of the project.

9.29 The Democratic Decision-Maker

The democratic decision-maker allows the team members to participate in the final decision. Voting by the group membership is critical and may even be mandatory. The company may even have a structured approach for this using guidelines or templates. This can happen even if the project manager is the expert in the area where the problem exists and even if the project manager has the authority to make the decision by himself/herself.

Democratic decision-making can create long-term problems. Team members may feel that they should be involved in all future decisions as well, even those where they may have limited knowledge about the problems. Asking team members to take an early vote on the solution to a problem can lead to apprehension if the team members are uncomfortable with making a decision based upon incomplete information. Waiting too long to make the decision can limit the options available and frustrate the project team because of the time that was wasted overthinking the problem and the solution.

Democratic decision-making is a strong motivational tool if used properly. As an example, if the project manager believes that he/she already knows the decision that should be made, asking the team for their opinion and giving credit to a team member for coming up with the same idea is a good approach. This encourages people to participate in decision-making and makes them believe that they will be given credit for their contributions.

9.30 The Self-Serving Decision-Maker

Everyone sooner or later is placed in a position where they must decide when making a decision what is more important, their individual values or the organizational values. This situation often forces people to make decisions either in favor of themselves or the organization. A compromise might be impossible.

These types of self-serving conflicts can permeate all levels of management. Executive may make decisions in the best interest of their pension rather than the best interest of their firm. One executive wanted to be remembered in history books as the pioneer of high-speed rapid transit. He came close to bankrupting his company in the process of achieving his personal ambitions at the expense of the projects he was sponsoring and at the expense of the corporation.

Self-serving decision-makers seem to focus on what is in their own best interest in the short-term and often disregard what might be in the best interest of the project. In a project environment, this can become quite a complex process if the team members, stakeholders, the client, and the project sponsor all want the decision made in their own best interest. Suboptimal solutions are reached with several parties being quite unhappy with the final result. Unfortunately, self-serving decisions are almost always made for what is in the best interest of the largest financial contributor to the project for fear that, if the financial contributor removes support from the project, the project may be canceled.

9.31 Delegation of a Decision-Making Authority

Effective managers, including many project managers, sometimes delegate decision-making authority and responsibility to subordinates when necessary. It is not uncommon in project management for many of the decisions to be made by stakeholders. An effective manager does not try to make every decision that is brought to them by subordinates, superiors, peers, or stakeholders. The prudent manager conserves his/her time and energy for those decisions which fall within his/her decision-making authority and responsibility. Every decision that is brought to a manager should be addressed in the context of: is this a decision I should make or does somebody else have the responsibility for making this decision within the organization?

It is important that the manager learn how to identify those decisions on problems/opportunities which he/she must make, those decisions which can be pushed up to the next higher level, and those decisions which have been delegated. Pushing up decisions to the superior should not be simply a passing of the buck, but it should be done in such a way that it represents a clear inability to make the decision at the manager's level, and a clear mandate for the decision to be made by the higher-level person.

Some guidelines that can be used to determine if the decision should be pushed up to the next higher level include answers to such questions as these:

- Are other departments of this organization or a stakeholder's organization affected?
- Will that decision have a significant impact on the supervisor's area of responsibility?
- Does it require information that is only available at higher levels?
- Does it involve a significant departure from existing business strategy?
- Does this problem rest within the delegation of authority and responsibility to this position?

9.32 Choice Elements of Decision-Making

Sometimes people make decisions on an informal assessment of alternatives. Such decisions are easy to make. Good decisions are more difficult. Each decision situation faced by a decision-maker contains four basic choice elements:

- What the decision-maker wants to do
- What can be done considering available resources
- What should be done to satisfy ethical and moral obligations
- What must be done to satisfy existing obligations such as allocating resources for pursuit of existing missions, objectives, goals, and strategies.

However rational the decision process becomes, the decision-maker will be influenced by these choice elements. Each choice based on an aspect of "can do," "must do," "want to do," and "should do," if taken by itself, could lead the organization in a different direction in its future. If an organizational decision-maker chooses a course of action based on personal desires, organizational ineffectiveness could result; for example, if a design engineer chooses a product design that he prefers, but which suboptimizes quality standards, the opportunity for product failure is enhanced. If a manufacturing manager neglects OSHA criteria in the design of production workstations, the opportunity for injury to the workers is increased. Selecting a course of action which does not support existing organizational strategies will create confusion in the organization. If the organization's capabilities cannot support a new strategic direction of the enterprise, then either the capabilities must be brought into line with the new strategic direction or that direction must change.

A decision-maker cannot separate the decision process from the organizational ambience where the decision is made and where it is executed. People who share resources with the decision-maker, if given the opportunity to participate in decision-making, can help to evaluate the "can do," "must do," "want to do," and "should do" elements of the decision.

There are limitations in the ability of humans to make purely objective decisions. No decision-maker ever has all the information with which to make a decision. Rather the decisions are made based upon "bounded rationality"; that is, the decision-maker collects relevant information and makes the decision within the boundaries established by that information. Another limitation facing the decision-maker is simply the inability to gather information directly. Rather the decision information is based upon what has been submitted by many others. The higher up in the organization the decision-maker happens to be, the more that decision-maker becomes dependent upon underlings for the information.

The decision-maker makes the decision-based in part upon his own self-interest and toward off any challenges to his decision-making authority. Information that a manager needs for making decisions, may very well be hidden, might not be relayed, or may be altered in such a way that it is distorted. Political behavior on the part of. People will often result in "satisficing" rather than maximizing benefits from the decisions. The decision-maker always is influenced by his values and experience background.

It is impossible to separate the decision-maker from how that person feels the decision will affect his or her own career. Given a choice, a decision-maker will make a decision based upon what would most favor his/her continued position within the organization.

9.33 Decision-Making Challenges

Success in managing project teams requires such obvious factors as top-management support, adequate role definition, and defined task boundaries. Feedback is also required so the teams know how well they can accomplish the things they set out to do.

In managing teams, the need for participation in decision-making is clear. Team leaders and team members should be aware of some of the difficulties of team decision-making:

- Inappropriate membership on the team.
- Lack of relevant information for team analysis.
- Lack of interest or weak concern by team members.
- Perceptions of team members that their participation is meaningless and that the decision-maker will act without really listening to their recommendations.

- Danger of interpersonal strife developing during the discussions about the decision. If the team interactions confront people rather than problems, an unsavory cultural ambiance can develop which will adversely affect the quality of the decisions.
- Dominance by the team leader or other members of the team which inhibits the participation of less assertive individuals.
- Inadequate team leadership which fails to define issues and decision parameters, encourages analysis, summarizes progress (or lack of progress) or focuses team efforts and brings the decision to a closure point.
- The dangers of Groupthink, a deterioration of mental efficiency, reality testing, and moral judgment that results from ingroup pressures.

Groupthink can damage the rational decision process if the desire for consensus goes beyond the need for effective decisions.

9.34 Examples of Decision-Making Challenges

Sometimes, there are challenges in the decisions that must be made and quite often it is solely the project manager's responsibility for making the decision. The following situations are all factual and provide examples of decision-making.

The Work-Life Balance

Regardless how frequently executives tell you that your project is extremely important to the success of the company and apply pressure upon you, what you have waiting for you at home is more important than any project. The question of course is whether you should make a decision that may impact your family.

Several years ago, a project manager turned around a failing project and received an award for his success and the decisions he had to make. When asked how he did it, he stated that he was the third project manager to manage this project. The first project manager was fired. The second project manager was one of his close friends. He watched his friend removed from the building in an ambulance. He then stated that he would not want this to happen to him.

In his first meeting with his team, most of whom had been working excessively overtime and under pressure from management for better results, he wondered what decisions he could make for a turn-around situation.

He told the team to stop working overtime. He also told the team to stop working on his project for a couple of weeks and spend whatever available time they have enjoyed with their family.

The team could not believe what they just heard. The project manager spent the next two weeks getting to understand the project, together with the accompanying problems and risks. At the end of the two weeks, the entire project team showed up in his office and asked:

> What can we do to help you turn around this project?

The project was transformed from a potential disaster to a success, and the project manager received an award. During the award speech, the project manager stated his appreciation to the team for the effort they provided and that he could not have been successful without their help. The team was highly appreciative of working for a project manager that demonstrated empathy for their feelings and demonstrated the importance of maintaining the correct balance between work and home life.

Another Form of the Work-Life Balance

An important project was already a few weeks behind schedule. An earthen dam burst in a nearby city causing the flooding of homes. Several of the project manager's team members immediately requested the project manager to allow them to leave work for a week to help fellow church members in the nearby city dig out their belongings. This would be a decision that executives wanted the project manager to make.

The project manager knew that the project would get further behind schedule but also believed that the workers would take off from work the following week regardless of what the project manager said. The project manager informed the customer of what happened and made it clear that the project manager authorized and approved the workers doing this. The customer, of course, was not pleased with the decision.

The workers were gone for two weeks rather than just one week. When the workers returned, the entire team worked unpaid overtime during the next few weeks and on weekends, and the project was completed on schedule. The workers were appreciative of what the project manager had done and volunteered to work for this project manager again.

Sharing Credit

In a Midwest city, a company finally completed a highly successful project that a local newspaper was following. At project completion, the newspaper interviewed the two individuals that headed up the project, namely the project manager and assistant project manager. The pictures of these two individuals also appeared in the newspaper along with the interview.

Members of the project team were furious that no mention was made of the team that brought the project to a successful conclusion. The team members believed that a picture of the team rather than the two individuals would have been better. The team members believed that they were exploited and stated that they would never work for either of these two individuals again if given the choice.

The company perceived this as a problem for resolution and made a company decision that project management is a team effort and recognition should be given to the entire team in most cases.

Rewarding Individuals

Project managers rarely possess real authority over all of the members of the project team and may have no direct or indirect input in the performance review process of the team members. As such, without any wage and salary administration responsibility, it is often difficult to reward people deserving of some form of recognition.

A project team was struggling to finish a project and generate the business benefits and business value expected by the customer. One of the "unionized blue-collar" workers believed that he could significantly increase the chances of success with added effort. For several months, he worked unpaid overtime. The project was finally completed, and the customer was delighted with the results.

Everyone in the company knew that the success was due mainly to the efforts of this one individual. The project manager wanted to reward the individual somehow but was unable to do so because of the union contract that specified the criteria for promotion and salary increases.

Undeterred and challenged, the project manager went to the president seeking advice on how to recognize the exceptional performance of a team member. The president held a meeting that included the entire team. He publicly praised the worker, handed him a company credit card, and told him to take his family on a six-day vacation in the Caribbean at the company's expense.

The president of the workers' labor union sent a letter to the company president commending him for the way he recognized and thanked a worker as being a significant contributor to the success of the company. The labor union also commended the project manager for the effort he exhibited in finding a way to recognize blue-collar worker performance. Workers were now volunteering to be assigned to this project manager's projects.

The Risks of Excessive Overtime

Overtime is a way of life on many projects and can impact how problems are solved and decisions are made. Excessive overtime can lead to costly mistakes resulting in schedule slippages as well as financial issues. During overtime on a large project, one of the team members used the wrong raw materials to build a product for testing. The results were disastrous and costly.

Senior management convened a meeting with the entire team and asked for the name of the individual responsible for the costly mistake. Before anyone could answer, the project manager spoke up and said that she was taking full responsibility for what happened by forcing people to work excessive overtime. Senior management fully understood what the project manager was trying to do with her statement and agreed to leave the meeting rather than looking for someone to potentially fire. The workers appreciated the project manager's actions and empathy for the workers.

Challenging Workers

Project managers often struggle when things are going bad and do not know to encourage team members or where to look for help in making a decision. Sometimes, the best support can come from the functional managers who assigned the team members to your project. As an example, a team member was struggling to find a solution to a technical problem. The project manager asked the functional manager if he could in some way assist in finding a solution to the technical issue.

The functional manager called a meeting with the project manager and the worker. The functional manager looked at the worker and said:

> You are one of my best employees and I know you are struggling to find a solution to the problem. I believed you were the best person in the department to handle this assignment. Since you cannot solve the problem, who do you think I should replace you with from our department that might have a better chance of solving the problem?

The worker now felt more challenged than ever before and asked for a bit more time. Eventually, the worker solved the problem, and the project was completed.

Discussion Questions

1. Does decision-making generally take more or less time than problem solving, and why?
2. Who determines if the attendees for decision-making are different than the attendees for problem-solving? What factors should be considered?
3. What are some of the different ways that a decision can be made?
4. Can the project or corporate culture impact how decisions are made? If so, how?
5. Why are there many different forms of decision-making? Can criteria for selecting an approach be identified?

6. Can several decision-making meetings be necessary before a final decision is made?
7. Who resolved conflicts that exist in decision-making meetings?
8. What are some of the advantages and disadvantages of group decision-making?
9. What are examples of mental roadblocks that affect decision-making?
10. Can you identify if someone is making a decision based entirely upon what is in their own best interest?
11. What are some of the challenges in setting up a decision-making meeting?

References

Cleland, D. I. and Kerzner, H. (1986). *Engineering Team Management.* New York: Van Nostrand Reinhold, pp. 230–234..

March, J.G. and Simon, H.A. (1958). *Organizations.* New York: Wiley.

Simon, H.A. (1957). *Models of Man: Social and Rational.* New York: Wiley.

Simon, H.A. (1977). *The New Science of Management Decisions,* rev. ed., Englewood Cliffs, N.J.: Prentice-Hall, p. 48.

10

Decision-Making Tools

10.0 Decision-Making Tools in Everyday Life

We all make decisions every some. We even use decision-making tools and do not realize it. Some of the decision-making tools and techniques people use in everyday life include:

- Determining the pros and cons for a given situation
- Choosing the alternative with the highest probability of occurrence
- Choosing the alternative that provides the greatest financial reward
- Choosing the alternative that offers the least amount of damage should something go wrong
- Accepting the first option that seems like it might achieve the desired result
- Following the advice of a subject matter expert
- Flipping a coin, cutting a deck of playing cards, and other random or coincidence methods
- Prayer, tarot cards, astrology, augurs, revelation, or other forms of divination

While these tools seem simplistic, there are more sophisticated tools that people use.

10.1 Use of Operations Research and Management Science Models

The evaluation of alternatives is an important step in the decision-making process. Quantitative techniques provide a basis for better assessment of alternatives through providing for the development and use of models to evaluate risk, uncertainty, and complexity in decision problems. Quantitative analysis provides a rational

Project Based Problem Solving and Decision Making: A Guide for Project Managers,
First Edition. Harold Kerzner.

Companion Website: www.wiley.com/go/kerzner/projectbasedproblemsolving

approach in the systematic application of scientific methods to the solution of management decisions. Quantitative techniques offer an alternative to intuitive "seat-of-the-pants" decision-making; these techniques provide the opportunity for the decision-maker to sharpen his judgment in making decisions. The final decision is always based on judgment.

Management science is often viewed solely as a mathematical or quantitative approach to decision-making. It is much more: a totality of many philosophies and techniques used in a systematic way to help the decision-maker make decisions.

10.2 SWOT Analysis

There are numerous complex tools that can be used for more complex decision-making tools. SWOT analysis looks at the *s*trengths, *w*eaknesses, *o*pportunities, and *t*hreats in a given situation. SWOT analysis was originally created as a strategic planning tool but has now been adapted to complex problem-solving on a project or in a business venture. It involves specifying the objective of the problem or project and identifying the internal and external factors that are favorable and unfavorable to achieve that objective.

A SWOT analysis must first start with defining a desired end state or objective for the problem at hand.

- **Strengths:** characteristics of the team that give it the ability to solve the problem. This could include technical knowhow and expertise.
- **Weaknesses:** characteristics that may prevent the team from solving the problem. This could be the team's lack of technical ability.
- **Opportunities:** *external* opportunities if the problem is resolved.
- **Threats:** *external* risks or elements in the environment or with stakeholders that could cause trouble for the project or business.

Strengths and weaknesses are internal strengths and weaknesses and look at the capability of the internal resources to solve the problem. Opportunities and threats are external results that may occur if the problem is or is not resolved. Strengths and weaknesses indicated what you "can do," and this must take place before you look at the opportunities and threats that indicate what you "should do." Having a great alternative that will appease the stakeholders is nice as long as you have the qualified resources to accomplish it.

10.3 Pareto Analysis

Pareto analysis is a statistical technique in decision-making that is used for selection of a limited number of tasks that produce significant overall effect such as the solution to a problem. It uses the Pareto principle – the idea that by doing 20% of work, 80% of the advantage of doing the entire job can be generated. In terms of quality improvement, a large majority of problems (80%) are produced by a few key causes (20%). In problem-solving, 80% of the desired solution can be obtained by performing 20% of the work.

Pareto analysis is a formal technique useful where many possible courses of action are competing for the solution to a problem. In essence, the problem solver estimates the benefit delivered by each action, then selects a number of the most effective actions that deliver a total benefit reasonably close to the maximal possible one.

Pareto analysis is a creative way of looking at causes of problems because it helps stimulate thinking and organize thoughts. However, it can be limited by its exclusion of possibly important problems which may be small initially, but which grow with time. It should be combined with other analytical tools such as failure mode and effects analysis and fault tree analysis for example.

This technique helps to identify the top 20% of causes of a problem that needs to be addressed to resolve the 80% of the problems. Once the top 20% of the causes are identified, then tools like the Ishikawa diagram or Fish-bone Analysis can be used to identify the root causes of the problems.

10.4 Multiple Criteria Decision Analysis

Multiple Criteria Decision Analysis (MCDA) or Multi-Criteria Decision-Making (MCDM) is a discipline aimed at supporting decision-makers faced with making decisions involving numerous and sometimes conflicting criteria. The conflicting criteria might be to lower cost, compress the schedule, improve quality, and minimize the risks on the project. Solving a problem that satisfies all the criteria may be impossible. MCDA aims at highlighting these conflicts during problem-solving and deriving a way to come to a compromise. It is a combination of intuition and a systematic approach.

When project teams are unable to satisfy all the criteria, decisions may end up being made based upon intuition or by selecting a preferred alternative. The role of the decision-making team might be to narrow down the choices to a small set of good alternatives before making a decision. This technique has been used on complex problems and generally leads to more informed and better decisions.

Unlike methods that assume the availability of measurements, measurements in MCDA are derived or interpreted subjectively as indicators of the strength of various preferences. Preferences differ from decision-maker to decision-maker, so the outcome depends on who is making the decision and what their goals and preferences are. Since MCDA involves a certain element of subjectivity, and the morals and ethics of the persons implementing MCDA play a significant part in the accuracy and fairness of MCDA's conclusions. The ethical point is very important when one is making a decision that seriously impacts other people, as opposed to a personal decision.

There are many MCDA/MCDM methods in use today. However, often different methods may yield different results for exactly the same problem. In other words, when exactly the same problem data are used with different MCDA/MCDM methods, such methods may recommend different solutions even for very simple problems (i.e. ones with very few alternatives and criteria). The choice of which model is most appropriate depends on the problem at hand and may be to some extent dependent on which model the decision-maker is most comfortable with. A question with all the above methods, and also methods not included in this list or even future methods, is how to assess their effectiveness.

10.5 Paired Comparison Analysis

In paired comparison analysis, also known as paired choice analysis, decision alternatives are compared two at a time to see the relative importance. The alternatives are compared, and the results are tallied to find an overall winner. The process begins by first identifying a range of plausible options. Each option is compared against each of the other options, determining the preferred option in each case. The results are tallied and the option with the highest score is the preferred option. This technique may be conducted individually or in groups. It may include criteria to guide the comparisons or be based on intuition following an open discussion of the group. A paired choice matrix or paired comparison matrix can be constructed to help with this type of analysis.

10.6 Decision Trees

A decision tree is a decision support tool that uses a tree-like graph or model of decisions and their possible consequences, including chance event outcomes, resource costs, and utility. It is one way to display an algorithm for decision-making. Decision trees are excellent tools for helping you to choose between several courses of action. They provide a highly effective structure within which you can lay out options and investigate the possible outcomes of choosing those options. They also help you to form a balanced picture of the risks and rewards associated with each possible course of action.

Decision trees are commonly used in operations research, specifically in decision analysis, to help identify the optimum approach most likely to reach a goal.

Amongst decision support tools, decision trees (and influence diagrams) have several advantages:

- Are simple to understand and interpret. People are able to understand decision tree models after a brief explanation.
- Lay out the problem such that all options can be analyzed.
- Allow us to see the results of making a decision.
- Have value even with little hard data. Important insights can be generated based on experts describing a situation (its alternatives, probabilities, and costs) and their preferences for outcomes.
- Use a white box model. If a given result is provided by a model, the explanation for the result is easily replicated by simple math.
- Can be combined with other decision techniques such as probability trees.

10.7 Influence Diagrams

An influence diagram (ID) (also called a relevance diagram, decision diagram, or a decision network) is a compact graphical and mathematical representation of a decision situation. It is a simple visual representation of a decision problem. Influence diagrams offer an intuitive way to identify and display the essential elements, including decisions, uncertainties, and objectives, and how they influence each other. It is a generalization of a Bayesian network, in which not only probabilistic inference problems but also decision-making problems (following maximum expected utility criterion) can be modeled and solved. Influence diagrams are very useful in showing the structure of the domain, i.e. the structure of the decision problem. Influence diagrams contain four types of nodes (*decision*, *chance*, *deterministic*, and *value*) and two types of arcs (influences and informational arcs).

As a graphical aid to decision-making under uncertainty, it depicts what is known or unknown at the time of making a choice, and the degree of dependence or independence (influence) of each variable on other variables and choices. It represents the cause-and-effect (causal) relationships of a phenomenon or situation in a non-ambiguous manner and helps in a shared understanding of the key issues.

10.8 Affinity Diagrams

An affinity diagram is a technique for organizing verbal information into a visual pattern. An affinity diagram starts with specific ideas and helps you work toward broad categories. This is the opposite of a cause-and-effect diagram, which starts with the broad causes and works toward specifics. You can use either technique to explore all aspects of an issue. Affinity diagrams can help you:

- Organize and give structure to a list of factors that contribute to a problem.
- Identify key areas where improvement is most needed.

This technique is useful when there are large amounts of data. The affinity diagram is a business tool used to organize ideas and data. The tool is commonly used within project management and allows large numbers of ideas stemming from brainstorming to be sorted into groups for review and analysis.

The benefits include:

- Adding structure to a large or complicated issue
- Breaking down a complicated problem into broad categories
- Gaining agreement on the solution to a problem.

10.9 Game Theory

Game theory models or games, as applied to project management problem-solving and decision-making, allow us to address a problem in which an individual's success in making choices depends on the choices of others. Simply stated, this technique considers responses of outside participants. It can be used to address how the client and stakeholders might react to certain alternatives selected.

It is used not only in project management, but also in the social sciences (most notably in economics, management, operations research, political science, and social psychology) as well as in other formal sciences (logic, computer science, and statistics) and biology (particularly evolutionary biology and ecology). While initially developed to analyze competitions in which one individual does better at another's expense (zero-sum games), it has been expanded to treat a wide class of interactions, which are classified according to several criteria. This makes it applicable to project management, especially on complex projects where multiple stakeholders exist, each with competing needs.

10.10 Cost-Benefit Analysis

Cost-benefit analyses are most useful for problems involving financial decisions. The alternatives to a problem are usually those where the value of receiving the benefits outweighs the costs of obtaining them. Factors considered in cost-benefit analyses include:

- Return on investment
- Net present value

- Internal rate of return
- Cash flow
- Payback period
- Market share.

Other parameters to consider that are more difficult to quantify include:

- Stockholder and stakeholder satisfaction
- Customer satisfaction
- Employee retention
- Brand loyalty
- Time-to-market
- Business relationships
- Safety
- Reliability
- Reputation
- Goodwill
- Image.

10.11 Nominal Work Groups

Work groups or nominal work groups, as applied to project management, can be an interdisciplinary collaboration of researchers or subject matter experts that have convened to identify and/or solve a problem. The group may be external consultants or contractors. The lifespan of the working group can be one day or several weeks. Such groups have the tendency to develop a *quasi-permanent existence* once the assigned task is accomplished; hence, the need to disband (or phase out) the work group once it has provided solutions to the issues for which it was *initially* convened.

The work group may assemble experts (and future experts) on a topic together for intensive work. It is not an avenue for briefing novices about the subject matter. Occasionally, a group might admit a person with little experience and a lot of enthusiasm. However, such participants should be present as observers and in the minority.

It is imperative for the participants to appreciate and understand that the working group is intended to be a forum for cooperation and participation. Participants represent the interests and views of stakeholders from disparate sectors of the community which happen to have a vested interest in the solution to a problem. Therefore, maintaining and strengthening communication lines with all parties involved is essential (this responsibility cuts both ways – stakeholders are expected to share what information, knowledge, and expertise they have on the issue.)

Each member of the work group may be asked to present their solutions to the rest of the group for analysis and be willing to accept constructive criticism. Work groups often have the advantage of arriving at a reasonably rapid decision but also suffer from the drawback that possibly not all of the alternatives were considered.

10.12 Delphi Technique

The Delphi technique is a structured communication approach, originally developed as a systematic, interactive forecasting and problem-solving method which relies on a panel of experts. The experts may not know who else is a member of the panel and all responses are provided anonymously.

In the standard version, the experts answer questionnaires in two or more rounds. After each round, a facilitator provides an anonymous summary of the experts' forecasts (or solutions to a problem) from the previous round as well as the reasons they provided for their judgments. Thus, experts are encouraged to revise their earlier answers in light of the replies of other members of their panel. It is believed that during this process the range of the answers will decrease, and the group will converge towards the "correct" answer. If convergence does not take place, the panel may be asked to select the five best alternatives for the next round. Then in the next round, select the three best alternatives. Then in the following round, select the two best alternatives. Finally, the process is stopped after a pre-defined stop criterion (e.g. number of rounds, achievement of consensus, and stability of results) and the mean or median scores of the final rounds determine the results.

Delphi is based on the principle that forecasts (or decisions) from a structured group of individuals are more accurate than those from unstructured groups. This has been indicated by the term "collective intelligence." The advantage of this technique is that the participants will provide their answers without being biased by others or openly criticized. People are free to state their opinion. The downside factors are that the process takes time (which may not be a luxury that most projects have) and that the best approach may be the combining of two or more alternatives rather than forcing people to select just one alternative.

10.13 Other Decision-Making Tools

There are other decision-making tools some of which may take longer to perform. These include:

- **Linear programming applications:** this includes the application of management science and operations research models for decision-making.
- **Trial and error solutions:** useful for small problems when the cause-and-effect relationships are reasonably well known.

- **Heuristic solutions:** similar to trial-and-error solutions but experimentation is done to reduce the list of alternatives.
- **Scientific methods:** this is used for problem-solving involving scientific issues where additional experimentation may be done to confirm the problem and/or hypothesis.

Problem-solving sessions normally involve only one decision-making tool. Most of the more complex tools are time-consuming and costly to use and combining several of these together may be prohibitive.

10.14 Artificial Intelligence

In the past five years, there have been numerous articles discussing how Artificial Intelligence (AI) can and will benefit the field of project management. As technology increases within the next two decades, AI is expected to replace humans in many of the simple and mundane tasks that are part of project and program management activities (Grace et al. 2018). The most common applications of AI are expected to be part of estimating and controlling project cost and time as well as resource management by determining the strength of employee qualifications for assignment to selected project activities. Other applications of AI are expected to include risk reduction practices, improved monitoring and control, status reporting, identification of anomalies, and correlations between projects (Ong and Uddin 2020).

Applications of AI already exist in today's businesses and industries. Examples of AI use include Alexa, Siri, Amazon's product recommendations, dating apps, and fitness trackers (Marr 2020; Dalcher 2022).

The growth of AI can be attributed to several factors, especially:

- Advances in technology
- Advances in software development
- Growth in information warehouses and business intelligence systems
- Growth in predictive analytics.

As the growth continues, disruptions are expected in the business models that companies use. The project management landscape will change significantly. Companies that are unwilling or unable to adapt to the changes generated by AI practices may not survive.

Recognizing the need for AI applications has been discussed in numerous journal articles. Many of the articles focus on the benefits expected and areas of applicability but spend little time discussing the challenges that companies will face. The intent of this section is to discuss the issues and challenges that may prevent successful AI implementation, at least in the short term, as part of project problem-solving and decision-making.

Customer Reaction

Gaining support for the implementation of AI practices should be significantly easier when the deliverables and outcomes of projects are for internal customers. External customers may not be supportive of the influence and decisions that AI software might have on their projects.

During competitive bidding practices, contractors may be required to include in their proposals how they plan on using AI during the execution of the customer's projects. Customers are interested in not only the final deliverables but the methodology used to achieve the results. Some customers may be hesitant to allow AI to make certain decisions on their projects. In the short term, contractors may find it necessary to have two forms of project management: traditional project management practices and AI project management practices. Customers may be given a choice of which project management approach they prefer.

In the past, customers were often prevented from participating in many of the decisions made by the contractors. As customers became more knowledgeable in project management, customer involvement was seen as beneficial and welcomed. Some customers may be unhappy that their involvement in projects is now being replaced by AI practices. Contractors using AI practices may find it necessary to educate customers on how AI will impact their projects.

Project Management Maturity

Project management maturity is usually defined in terms of continuous improvement efforts in the forms, guidelines, processes, templates, and checklists used in project management. About two decades ago, organizations and professional societies such as PMI® created project management maturity models (PMMMs) to assist organizations in their quest for project management maturity.

Most maturity models were based upon questionnaires or assessment instruments related to how well the company was aligned to the processes, knowledge areas, and domain areas identified in PMI®'s ***PMBOK® Guide***. The PMMMs will now have to go through continuous revisions to allow for the impact of AI practices on project management maturity and the ability of AI practices to make some critical decisions. Not all knowledge areas are expected to undergo AI transformations at the same time. The easiest knowledge areas for including AI practices might be time, cost, and integration management. Other knowledge areas may be more challenging in accepting AI practices even if they are connected to areas that are heavy users of AI techniques.

Educating Project Managers

There are numerous universities and private sector companies offering training programs for project managers. Training normally centers around the knowledge and domain areas in the ***PMBOK® Guide*** and the ***Standard for Project Management***. Unfortunately, as of now, concepts related to AI are not included.

As technology replaces the simpler or mundane project and program tasks, project managers can be expected to be integrated into the higher levels of project management such as in fuzzy front-end project selection and prioritization processes, strategic planning, and strategic business decision-making as it relates to projects.

For an understanding of AI concepts and how they are used in project decision-making, PMs may require training that include tools related to statistics, software development, advanced data analyses, trend analyses, and data intuition analyses. The tools are expected to be different for each life cycle stage of a project since the problems and necessary decisions can be based upon the position in the project's life cycle.

The educational modules related to AI will be based upon information and data collected from a variety of internal and external sources but intended for use in the parent company. As such, the use of trainers external to the company for AI education may be restricted to general AI knowledge training. There may be many customized company software tools and because the data may include company-sensitive information, how the tools should be used based upon the information in the knowledge repository may require internal training programs possibly conducted by one of the parent company's project management offices (PMOs).

Educating Team Members

For almost five decades, project management practices have used documents such as the LRC (Linear Responsibility Chart) and the RACI Chart (responsibility, accountability, consulted, and informed) to assign responsibilities to team members. The charts were filled in at the onset of the project and updated as necessary as the project progresses.

By retaining the information in the charts, companies can use AI practices to assign the best-qualified resources to project activities based upon historical data. The goal is to find the right skills for the right job and to identify a resource shortage if one exists. Additional information other than skills that may be collected might include employee historical performance review data, education, training programs attended, employees/stakeholders they have worked with in the past, and types of project assignments.

While this sounds like a good idea, team members may not be as open-minded as project managers when it comes to accepting AI as part of their job. Two of the concerns that employees will face include: (i) people may be afraid that AI will take away jobs, and (ii) people fear failure (Wang 2019). Employees believe that certain assignments may provide them with more opportunities for career advancement and may resist assignments made by AI. Employees may not be given a choice of assignments as they had been given in the past. Another concern is that AI may end up making decisions that employees made in the past and their jobs may no longer be required. Finally, AI may make decisions that the employees might not have made. If the decision turns out to have an unfavorable impact on the project, the employee may see this as a reflection on his/her ability.

Employees will be required to use AI technologies as part of their job. This will need to be integrated into all training courses to gain support and effective use of the AI software and data.

10.15 Risk Management

Project teams have great expectations that AI practices will help them reduce project risks by analyzing past data and running multiple scenarios to determine realistic outcomes and the lowest risk path. The expectation is better risk management decision-making. Unfortunately, there are several challenges that must be overcome before this becomes a reality. All projects today seem to be impacted by the VUCA (volatility, uncertainty, complexity, and ambiguity) environment such that many estimates and forecasts are at best intuition. Reducing uncertainty may be achievable but not a complete elimination of uncertain events.

Some degree of "humanness" must exist in all risk analyses activities. Even if an abundance of data exists in the information warehouse, the necessity for involvement and understanding of the decision-making process by humans will be necessary. AI optimization of tasks will not replace human involvement. Project teams still need behavioral skills such as conflict resolution, emotional intelligence, persuasion, team building, and effective collaboration.

Projects are managed by people, not tools. Regardless of how much technology will be included in project management, the human touch of collaboration and interaction among team members and stakeholders will not disappear. As stated by Gupta (2020):

> We cannot discount the human touch aspect when we are dealing with humans. Humans are social animals. We want to work with each other [rather than just with machines].

The human side or project management is also needed to evaluate if the decision made by AI is realistic. The project team must determine if the decision made by AI was based upon:

- The wrong information
- Partial information
- Conflicting information
- The correct past experiences.

Ethics

AI is expected to provide the project management community of practice with more possible outcomes and the opportunity for better decision-making. The complex analytics track performance, look for trends, and make assumptions about where the project is heading. This assumes that the data inputted into the system was not prone to human error and that the outcomes were calculated in an ethical manner. As stated by Dalcher (2022):

> Humans struggle to acknowledge and communicate tacit knowledge. While machines are more capable at capturing the aspects that we cannot articulate by observing and deducing patterns, we are often at a loss to figure out how a machine system has reached particular decisions.

AI systems have difficulty understanding biases and the recommendations made because of the biases. Brynjolfsson and McAfee (2017) discuss hidden biases that can exist in the data and the difficulty in understanding and correcting errors.

Data gathering and data analytics are always prone to human error. AI cannot tell if critical data is missing. All AI models have a need for massive amounts of information. The data should be diversified and from multiple sources. Having data from a single source is an invitation for poor decision-making.

On projects related to R&D, innovation, and new product development, limited information may exist for reliable AI predictions. Also, there may exist multiple definitions of success that could lead AI analytics in the wrong direction. As an example, a company may be willing to invest significant funds into a portfolio of projects designed to operate at a loss initially to penetrate a new market. AI may be unfamiliar with the scope and misinterpret the real definition of success. This could result in faulty conclusions from AI software. AI is prone to errors especially if the database contains corrupt information. Legally, AI software cannot be held accountable if the wrong decisions are made and leads to bad consequences.

Another issue with AI ethics relates to the security of the information in the database. Security standards must be established on how the secured information is used and who may have access to the information without violating the right to privacy.

AI Penetration

AI penetration is expected to be part of all project management practices over the next 20–30 years and will occur slowly because of the potential for significant resistance. PMOs will most likely take the lead and help introduce AI practices in the project management areas of knowledge that might provide the least degree of resistance. Later, as company-wide training programs are launched focusing on the use of AI practices, acceptance should be a bit easier once employees recognize the impact that AI can have on decision-making, reporting, forecasting, predictive analysis, and resource allocation.

There is no question that AI is gaining popularity and will become part of every company's continuous improvement effort to maintain its competitive position. The time saved by allowing AI to assist in repetitive or routine tasks can be redeployed to other activities. However, implementation of AI is not free of risks and challenges. Knowing when and how to deploy AI practices are critical to avoid abuse of AI systems and to gain employee acceptance.

Discussion Questions

1. When should operations research or management science models be used for decision-making compared to other type of models?
2. On what type of project problems should SWOT analysis be used rather than other approaches?
3. On what type of project problems should a cost-benefit analysis be used rather than other approaches?
4. What are some of the risks in allowing artificial intelligence to make some decisions on projects?

References

Brynjolfsson, E. and McAfee, A. (2017). The business of artificial intelligence: how AI fits into your data science team. *Harvard Business Review* 98 (4): 3–11.

Dalcher, D. (2022). The quest for artificial intelligence in projects. *Advances in Project Management Series, PM World Journal* XI (III).

Grace, K., Salvatier, J., Dafoe, A. et al. (2018). When will AI exceed human performance? Evidence from AI experts. *Journal of Artificial Intelligence Research* 62: 729–754.

Gupta, C. (2020). Artificial Intelligence (AI) influence in Project Management. *PM World Journal* IX (II).

Marr, B. (2020). *Tech Trends in Practice: The 25 Technologies That are Driving the 4th Industrial Revolution*. Chichester: Wiley.

Ong, S. and Uddin, S. (2020). Data Science and Artificial Intelligence in Project Management: the past, present and future. *Journal of Modern Project Management* 7 (4): 1–8.

Wang, Q. (2019). How to apply AI technology in Project Management. *PM World Journal* VIII (III).

11

Predicting the Impact

11.0 Evaluating the Impact of a Decision

Anybody can make a decision, but the hard part is making the right decision. Decision-makers often lack the skills in how to evaluate the results or impact of a decision. What the project manager believes was the correct decision may be viewed differently by the client and the stakeholders.

Part of decision making requires the project manager to predict how those impacted by the decision will react. Soliciting feedback prior to the implementation of the solution seems nice to do. But the real impact of the decision may not be known until after full implementation of the solution. As an example, as part of developing a new product, marketing informs the project manager that the competition has just come out with a similar product and marketing believes that we must add in some additional features into the product you are developing. The project team adds in a significant number of "bells and whistles" to the point where the product's selling price is higher than that of the competition and the payback period is now elongated. When the product was eventually launched, the consumer did not believe that the added features were worth the additional cost.

It is not always possible to evaluate or predict the impact of a decision when making a choice among alternatives. But soliciting feedback prior to full implementation is helpful.

11.1 Creating a Consequence Table

A useful tool for assisting in the selection of alternatives is a consequence table as shown in Exhibit 11.1. For each alternative, the consequences are measured against a variety of factors such as each of the competing constraints. For example, an

Project Based Problem Solving and Decision Making: A Guide for Project Managers,
First Edition. Harold Kerzner.

Companion Website: www.wiley.com/go/kerzner/projectbasedproblemsolving

Competing Constraints

Alternative	Time	Cost	Quality	Safety	Overall Impact
#1	A	C	B	B	B
#2	A	C	A	C	B
#3	A	C	C	C	C
#4	B	A	C	A	B
#5	A	B	A	A	A

A = High Impact
B = Moderate Impact
C = Low Impact

Exhibit 11.1 Creating a Consequence Table

alternative could have a favorable consequence on quality but an unfavorable consequence on time and cost. Most consequence tables have the impacts identified quantitatively rather than qualitatively. Risk is also a factor that is considered, but the impact of risk is usually defined qualitatively rather than quantitatively.

If there are three alternatives and five constraints, then there may be 15 rows in the consequence table. Once all 15 consequences are identified, they are ranked. They may be ranked according to either favorable or unfavorable consequences. If none of the consequences are acceptable, then it may be necessary to perform tradeoffs on the alternatives. This could become an iterative process until an agreed-upon alternative is found.

The people preparing the table are the people that make up the project team rather than possible outsiders that were brought in as subject matter experts for a particular problem. Project team members know the estimating techniques as well as the tools that are part of the organization process assets that can be used for determining impacts.

11.2 Performing Impact Analysis

It is nice to have several possible alternatives for the solution to a problem. Unfortunately, the alternative that is finally selected must be implemented and that can create problems also.

One of the ways to analyze the impact is to create an impact-implementation matrix as shown in Exhibit 11.2. Each alternative considered could have a high or low impact on the project. Likewise, the implementation of each alternative could be easy or hard.

Each alternative is identified in its appropriate quadrant. The most obvious choice would be the alternatives that have a low impact and are easy to implement. But in reality, we often do not find very many alternatives in this quadrant.

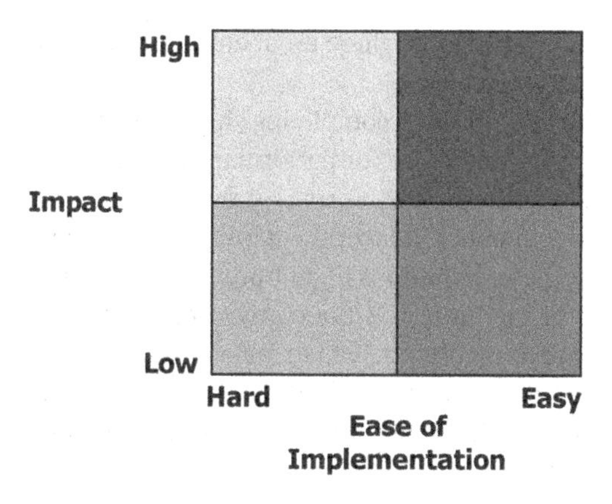

Exhibit 11.2 Performing Impact Analysis

11.3 The Time to Implement a Solution

Too many times in my life, I have sat in problem-solving sessions and listened to team members come up with good (often brilliant) solutions to a problem. Everyone becomes enamored with the brilliance of the solution, but nobody seems concerned as to how long it will take to implement the solution. Significantly more time is consumed in implementing a solution than in decision-making.

Questions that should be considered include:

- Must we change our plans/baselines and, if so, how long will it take?
- How long will it take to get the necessary additional funding approved?
- Will the resources with the required skill levels be available when needed?
- Is overtime an option?
- How long will it take to procure those materials we need?
- Are additional reviews and meetings necessary before implementation can begin?
- Are additional reviews and meetings necessary as we implement the solution?

Simply stated, decision making is easy; implementation is often difficult and time-consuming.

11.4 The Definitions for Project Success and Failure Are Changing

Most people have a relatively poor understanding of what is meant by project success and project failure. Project problem-solving and decision-making are designed to achieve project success. Unfortunately, there are many definitions of success, and this impacts how we resolve problems.

Project success has traditionally been defined as completing the requirements within the triple constraints of time, cost, and scope (or performance). This is the answer that had been expected of students on most exams. In the same breath, project failure had been defined as the inability to meet the requirements within time, cost, and scope. Unfortunately, these definitions do not provide a clear picture or understanding of the health of the project, if the correct decisions were made, and whether or not success has been achieved. And to make matters worse, the definition of success or failure is treated like the definition of beauty; it is in the eyes of the beholder. Today, we are finally beginning to scrutinize the definitions of project success and project failure.

Historical Perspective

The complexities with defining project success and failure can be traced back to the early days of project management. The birth and initial growth of project management began with the Department of Defense (DOD) in the United States. With thousands of contractors, DOD wanted some form of standardization with regard to project performance reporting. The Earned Value Measurement System (EVMS) was created primarily for this purpose.

For the EVMS to be effective, metrics were needed to track performance and measure or predict project success. Everybody knew that measuring success was complicated and that doing it correctly required several metrics. Unfortunately, our understanding of metrics and metric measurement techniques was relatively poor at that time. The result was the implementation of the rule of inversion. The rule of inversion states that the metrics with the highest information value, especially for decision-making and measuring success, should be avoided, or never measured, because of the difficulty in data collection. Metrics like time and cost are the easiest to measure and should therefore be used. The result was that we spent too much time on these variables that may have had the least impact on decision-making and measuring and predicting project success. The EVMS, for all practical purposes, had two and only two metrics: time and cost. Several formulas were developed as part of the EVMS, and they were all manipulations of time and cost.

The definition of success was now predicated heavily upon the information that came out of the EVMS, namely time and cost. The triple constraints of time, cost, and scope were established as the norm for measuring and predicting project success. Thus, the criteria for solving a problem focused heavily upon the triple constraints.

Unfortunately, good intentions often go astray even when we believe we have solved the problem and made the correct decisions. DOD's contracts with the

Aerospace and Defense Industry were heavily based upon the performance of the engineering community. In the eyes of the typical engineer, each of the triple constraints did not carry equal importance. For the engineers, scope and especially technical achievement were significantly more important than time or cost. DOD tried to reinforce the importance of time and cost, but as long as DOD was willing to pay for the cost overruns and allow schedule slippages, project success was measured by how well performance was achieved. To make matters worse, many of the engineers viewed project success as the ability to exceed rather than just meet specifications and to do it using DOD's funding. Even though the triple constraints were being promoted as the definition of success, performance actually became the single success criterion.

Early Modification to the Triple Constraints

DOD's willingness to tolerate schedule slippages and cost overruns for the sake of performance gave the project management community the opportunity to consider another constraint, namely customer acceptance. Projects, by definition, are most often unique opportunities that you may never have attempted before and may never attempt again. As such, having accurate databases that can be used to predict the time and cost to achieve success was wishful thinking. Projects that required a great deal of innovation were certainly susceptible to these issues. To make matters worse, the time and cost constraints were being established by people that knew very little about the complexities of project management and had never been involved in innovation activities.

People began to realize that meeting the time and cost constraints precisely would involve some degree of luck. Would the customer still be willing to accept the deliverables if the project was late by one week, two weeks, or three weeks? Would the customer still be willing to accept the deliverables if the cost overrun was $10,000, $20,000, or $100,000?

Many companies changed their view of success such that the only definition of success was now customer satisfaction or customer acceptance. For some customers and contractors, time and cost were insignificant compared to customer satisfaction. Having the deliverables late or over budget was certainly better than having no deliverables at all.

The Use of Primary and Secondary Constraints

As projects became more complex, organizations soon found that the triple constraints were insufficient to clearly define project success even if the constraints were prioritized. This posed challenges for problem-solving and decision-making. There were other constraints that were often more important than time, cost, and scope. These "other" constraints were referred to as secondary constraints with time, cost, and scope being regarded as the primary constraints. Typical secondary constraints included:

- Using the customer's name as reference at the completion of the project
- Probability of obtaining follow-on work
- Financial success (i.e. profit maximization)
- Achieving technical superiority (i.e. competitive advantage)
- Aesthetic value and usability
- Alignment with strategic planning objectives
- Meeting regulatory agency relationships
- Abiding by health and safety laws
- Maintaining environmental protection standards
- Enhancing the corporate reputation and image
- Meeting the personal needs of the employees (opportunities for advancement)
- Supporting and maintaining ethical conduct (Sarbannes–Oxley Law).

The secondary constraints created challenges for many companies and made problem-solving and decision-making difficult. The EVMS was created to track and report only the primary constraints. To solve the tracking problem, companies created enterprise project management methodologies (EPMs) that incorporated the EVMS and tracked and reported the secondary constraints. This was of critical importance for some companies because the secondary constraints could be more important than the primary constraints on some problems.

From Triple Constraints to Competing Constraints

When PMI released the 4th edition of the ***PMBOK® Guide***, the use of the term triple constraints was replaced with the term "competing constraints." Defining project success was now becoming significantly more complicated because of the increasing number of constraints and their importance in defining project success. Everybody knows and understands that "what gets measured, gets done." Therefore, there were three challenges that soon appeared:

- Each new constraint must be tracked the same way that we traditionally tracked time and cost.
- To track the new constraints, we need to establish metrics for each of the constraints. You cannot have a constraint without having a metric to confirm that the constraint is being met.
- Metrics are measurements. We must understand the various measurement techniques available for tracking the new metrics that will be used to predict and report success.

Project success, metrics, and measurement techniques were now interrelated. The definition of success now became a critical issue when looking at alternatives to solve problems.

11.5 Project Decision-Making and Politics

There are several factors that are changing the landscape for project management, many of which have been discussed earlier. Some of the factors include:

- We are now working on significantly more projects, especially strategic and innovation projects that have more unknowns.
- The direction of the strategic and innovation projects can change over the life cycle of the projects, thus increasing the number of decisions that must be made.
- The projects are longer than in the past.
- There is a greater likelihood that workers will be assigned full-time rather than part-time.
- We have a growth in the development of flexible methodologies and each project can have a different methodology and different decisions to be made
- Stakeholder involvement in project decision-making is increasing.

All six of these factors have a common characteristic: they make project decision-making more complex and require much greater collaboration with and between team members, stakeholders, and customers than in the past. Increases in the need for collaboration will most certainly be accompanied by an increase in project politics.

Many people believed that project politics were mainly restricted to multinational projects where we may have to deal with political instability in governments. The outcome of a political risk might be that a new political party takes control of the government and project funding is cut or there is a significant change in the direction of the project. Political risk could also alter the raw materials you must use, the sources of the materials (countries from which the materials come), and which contractors you must use.

On traditional projects, project politics were often misunderstood and avoided with the belief that it was self-serving and would lead to negative results. In the past, we relied heavily upon the project sponsor or the governance committee to resolve problems that were accompanied by the political risks. Today, the responsibility for political risk management falls on the shoulders of the project managers because many of the causes of political risk are the result of actions or requests made by the people empowered to provide project governance. Every project has some form of political issues. It can also come from customers as well as stakeholders trying to redirect the project because of personal agendas.

We now live in a VUCA environment where the impact of enterprise environmental factors (EEFs) is growing. Project political risks are no longer attributed just to multinational projects. Project politics is most frequently treated in project management as part of risk management practices as a political risk. The result of a political risk can alter the outcome of project decision-making by changing the probability of meeting the original business objectives.

The longer the project and the less information you have (i.e. such as on strategic or innovation projects) the greater the need to understand as quickly as possible if people have personal or political agendas. The more ambiguity and uncertainty you have, the greater the risk of political decision-making. When politics becomes important, people selecting the projects may not express all their real concerns for fear of exposing their true agenda.

Project managers must learn how to manage projects in a political environment. They must be willing to understand who the stakeholders and decision-makers are and develop a stakeholder management or political plan as part of the project's communications plan.

Discussion Questions

1. Why is it difficult to determine the impact of a project decision?
2. What is a consequence table and when should it be used?
3. How important is it to determine the time needed to implement a solution?
4. Why are there different definitions of project success and failure?
5. Can politics become more important in making a decision and selection of the best alternative? If so, how can this be overcome by the project manager?

12

The Need for Effective or Active Listening Skills

12.0 Active Listening

To be highly effective at problem-solving and decision-making, people must have good communication skills. Part of good communications skills is effective or active listening. Improper listening can result in miscommunication, numerous and costly mistakes, having to repeat work, schedule delays, and the creation of a poor working environment. The result of poor listening is often more team meetings than originally thought and an abundance of action items. This can happen in any of the Domain Areas of the ***PMBOK® Guide***.

Active listening techniques make it easier for the listener to understand, interpret, and evaluate what is being said. Active listening is essential during problem-solving and decision-making sessions because it can reduce conflicts, strengthen cooperation among the participants, and promote a better understanding of the issues.

When people interact, they often do not listen attentively. This may be the result of focusing on other issues rather than the speakers. People may be thinking about other things, having one's mind being judgmental of what is being said, wondering what they must say when it is their turn to speak, or thinking of excuses why they should not have to speak at all. Avoiding these mental activities is not an easy task.

12.1 Active Listening Body Language and Communications

Active listening involves more than just listening to the words that the speaker says. It also involves reading body language. Sometimes, the person's body

Project Based Problem Solving and Decision Making: A Guide for Project Managers,
First Edition. Harold Kerzner.

Companion Website: www.wiley.com/go/kerzner/projectbasedproblemsolving

language provides the listener with a much more accurate understanding of the intent of the message.

All elements of communication, including active listening, may be affected by barriers that can impede the flow of conversation. Such barriers include distractions, trigger words, poor choice of vocabulary, and limited attention span. Listening barriers may be psychological (e.g. emotions) or physical (e.g. noise and visual distraction). Cultural differences including speakers' accents, vocabulary, and misunderstandings due to cultural assumptions often obstruct the listening process. Frequently, the listener's personal interpretations, attitudes, biases, and prejudices lead to ineffective communication.

Although we often talk about the listener when we discuss active listening barriers, it should be noted that sometimes the barriers to active listening are created by the speaker. This can occur when the speaker continuously changes subjects, uses words and expressions that confuses the listener, distracts the listener with improper or unnecessary body language, and neglects to solicit feedback as to whether the listener truly understood the message.

12.2 Active Listening Barriers Created by the Speaker

Typical active listening barriers created by the speaker include:

- Creating a communications environment where excessive note-taking is required such that the listener never gets to digest the material or see the body language
- Allowing constant interruptions to take place which can lead to conflicts and arguments, or allowing the interruptions to get the actual subject of the communications way off track
- Allowing people to cut you off, change subjects and/or defend their positions
- Allowing for competitive interruptions
- Speaking in an environment where there may be excessive noise or distractions
- Talking without pauses or talking too fast
- Neglecting to paraphrase or summarize critical points
- Failing to solicit feedback by asking the right questions
- Answering questions with responses that are slightly off.

12.3 Active Listening Barriers Created by the Listener

Typical active listening barriers created by the listener include:

- Looking at distractions rather than focusing on the speaker

- Letting your mind wander and looking off in the distance rather than staying focused
- Failing to ask for clarification of information that you do not understand
- **Multitasking:** doing a task, such as reading, while the speaker is presenting his/her message
- Not trying to see the information through the eyes of the speaker
- Allowing your emotions to cloud your thinking and listening
- Being anxious for your turn to speak
- Being anxious for the meeting to be over.

12.4 Overcoming Active Listening Barriers

Active listening barriers can be overcome. To use the active listening technique to improve interpersonal communication, one puts personal emotions aside during the conversation, asks questions, and paraphrases back to the speaker to clarify understanding, and one also tries to overcome all types of environmental distractions. Don't judge or argue prematurely. Furthermore, the listener considers the speaker's background, both cultural and personal, to benefit as much as possible from the communication process. Eye contact and appropriate body languages are also helpful. It is important to focus on what the speaker is saying; at times, you might come across certain keywords which will certainly help you understand the speaker. The stress and intonation will also keep you active and away from distractions. Taking notes on the message will aid in retention.

12.5 Techniques for Effective Listening

Some techniques for active listening effectiveness might include:

- Always face the speaker
- Maintain eye contact
- Look at the speaker's body language
- Minimize distractions, whether internal or external
- Focus on what the speaker is saying without evaluating the message or defending your position
- Keep an open mind on what is being discussed and try to empathize with the speaker even if you disagree
- Do not interrupt the speaker even though you have a different position.

Discussion Questions

1. What is meant by active listening?
2. Active listening is a component of what major project management skill that project managers should possess.
3. What are some examples of barriers created by the speaker that prevent people from listening correctly?
4. What are some examples of barriers created by listeners that prevent them from effectively hearing what is being discussed?

13

Barriers

13.0 The Growth of Barriers

The Project Management Institute (PMI) recently celebrated its 50th anniversary. Even though most of the core concepts of project management have been recognized and used successfully for decades, there is still resistance in the form of barriers that can prevent successful implementation of all or specific components of project management. The barriers can create issues that make problem-solving and decision-making difficult, if not impossible. As new techniques begin being used in the project management environment, such as the impact of digitalization, artificial intelligence, the Internet of things (IOT), big data, blockchain, and disruptive project management practices, new barriers are expected to appear. An understanding of the barriers can help us prevent or diminish their impact.

Up until about 10 years ago, there appeared to be limited published research on the identification and impact of these barriers. Part of the problem was that the literature at that time seemed to focus heavily upon successes rather than failures because nobody wants to admit to having made a mistake. Today, we recognize that we may discover more opportunities for continuous improvement efforts from failures and mistakes rather than from best practices and lessons learned.

In an early paper by Kerzner and Zeitoun (2008), the authors focused heavily upon the barriers that existed primarily in emerging markets. The authors stated:

> Growth in computer technology and virtual teams has made the world smaller. Developed nations are flocking to emerging market nations to get access to the abundance of highly qualified and relatively inexpensive human capital who want to participate in virtual project management teams.
>
> A multi-national virtual project management team, however, may come with headaches. Because of the growth of project management worldwide, many

Project Based Problem Solving and Decision Making: A Guide for Project Managers, First Edition. Harold Kerzner.

Companion Website: www.wiley.com/go/kerzner/projectbasedproblemsolving

> executives openly provide lip service to its acceptance, yet behind the scenes, they erect meaningful barriers to prevent it from working properly. This creates significant hardships for those portions of the virtual team that must rely upon their team members in emerging market nations for support.
>
> Barriers to effective project management implementation exist worldwide, but in emerging market nations, the barriers are more apparent. To be aware of the possible barriers and their impact on project management implementation allows us proactively to begin to surmount them.

Today, many of the barriers that appeared previously in primarily emerging markets are now quite apparent in developed nations and within areas of companies that may have been using project management for decades. Barriers are no longer restricted just to specific countries or nations. Some barriers may be industry-specific, appear in certain functional disciplines of a company, or occur because of the personal whims of some managers and executives. Barriers can appear anywhere and at any time. And almost always, the barriers seem to prevent the best decisions from being made.

Some industries appear to be more prone to project management implementation barrier research than others. The barriers in the IT industry have been discussed in the literature by Terlizzi et al. (2016), Johansen and Gillard (2005), Khan et al. (2011), Khan and Keung (2016), Polak and Wójcik (2015), Carvalho (2014), and Niazi et al. (2010). Research and development barriers have been addressed by Santos et al. (2012), Sommer et al. (2014), and Sakellariou et al. (2013/2014). Recently, there has also been research in public sector barriers as described by Blixt and Kirytopoulos (2017).

Another industry commonly discussed is the construction industry as indicated by Arnold and Javernick-Will (2013), Hwang and Tan (2012), Senesi et al. (2015), Loushine et al. (2006), and Moore and Dainty (2001). Some authors focus on barriers in specific countries. As example, Tang et al. (2007) looked at the Chinese construction industry whereas Hwang et al. (2014) researched small construction projects in Singapore. Magnier-Watanabe and Benton (2013) examined barriers facing Japanese engineers.

There has also been research in barriers that can affect certain ***PMBOK® Guide*** Areas of Knowledge or specific project management processes, tools, and techniques. Kutsch and Hall (2009, 2010) and Paterson and Andrews (1995) looked at the barriers that impact risk management. Ambekar and Hudnurkar (2017) focused on the use of six sigma. Ali and Kidd (2014) examined configuration management activities and Hwang et al. (2017) investigated barriers affecting sustainability efforts.

There are numerous possibilities for classifying the different barriers affecting project management. This chapter briefly discusses some of the more prominent barriers in the categories identified in Exhibit 13.1 because of their impact on project problem-solving and decision-making.

Exhibit 13.1 Categories for Barriers

13.1 Lack of Concern for the Workers Barriers

Whenever we change or introduce new management processes, whether it is for project management, Agile, Scrum, six sigma, or other practices, we must consider the impact on the wage and salary administration program and the expectations the workers have as to the impact on their career. Workers expect to be recognized, or even rewarded, for good performance including the problems they solve and the decisions they made. Unfortunately, we often introduce or solve problems without considering employee performance review implications until some damage occurs and barriers form to prevent the changes from being implemented correctly. The result is usually poor decision-making.

Sometimes human resources barriers are created that can cause a conflict between what appears to be in the best interest of the project and the best interest of the worker. It is not uncommon for the project team to fail to realize the impact of the barrier or even that it exists until the project is over. In most cases, as shown in Exhibit 13.2, the result may be limited project success, or possibly even failure based upon when the barrier is recognized.

Situation 1 (The Co-Location Barrier): A project manager working for a government agency was placed in charge of a two-year project and wanted a co-located team. The PM was fearful that, if the team members were to remain in their functional areas, the functional managers might use the workers frequently on other projects thus impacting the schedule of his project. During project staffing, the PM also demanded the best resources knowing full well that many of the workers would be overqualified for the tasks and therefore underutilized. Although the demand for the best resources benefited this project manager's assignment, other projects which required workers with these specific skills were shorthanded and

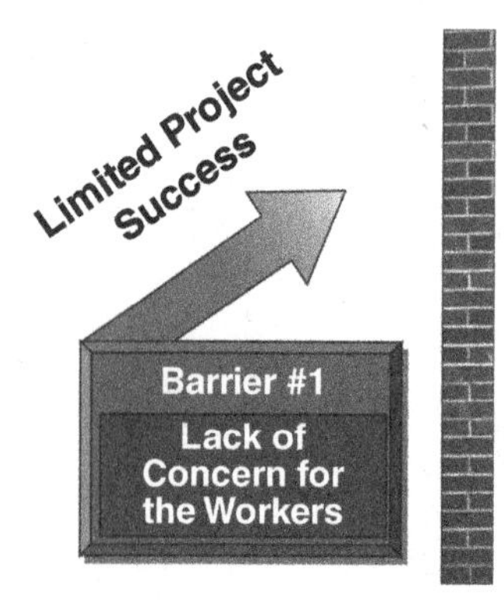

- The barrier may have a favorable or unfavorable impact on the performance of the project but creates problems for the worker's future career goals.

Exhibit 13.2 Human Resources Management Barrier

struggled. The project manager's decision may have benefited the project but not necessarily the entire company.

The project manager found a vacant floor in a government building and all the workers were relocated to this location on a full-time basis even though the assignment did not necessarily mandate full-time workers. However, even though the workers were removed from their functional organizations, their functional managers were still responsible for their performance reviews.

At the end of the two years, the project was deemed as a success. However, many of the workers were quite upset because:

- Most of the workers were given mediocre performance reviews during the two-year period because their functional managers were unaware of their performance.
- When given the choice of who deserved a promotion during the two-year period, the functional managers promoted those workers first that resided in the functional area and benefited from a multitude of functional area projects.
- Some of the workers discovered that their functional managers filled their vacated position with other workers and that these now displaced team members had to find positions elsewhere, and possibly lose some seniority.

In Situation 1, which occurred in a developed nation, the project was a success, and the project manager received a promotion. Unfortunately, the workers did not see any benefits to their career goals by working on this project and stated that they would not desire to work for this PM again. The organization had to rethink the benefits of using a co-located team approach. Barriers can exist anywhere. While the decision made by the project manager may have benefited the project and the project manager's personal agenda, the same decision caused havoc for the company and alienated the workers.

Situation 2 (The Prolonged Employment Barrier): A government-run utility in an emerging market nation embarked upon a three-year project to build a new power generation plant. To minimize the cost of the project and support the local economy, the decision was made to use local and country-wide labor rather than hiring more expensive contractors external to their country. This would have the additional benefit of providing employment for many local workers.

The workers were happy with the opportunity for employment but were fearful of what might happen when the project comes to an end. To guarantee long-term employment, and possibly retirement benefits, the workers began slowing down the project, made decisions that required rework, and began making mistakes, to the point where the schedule was now stretched out to 10 years. The organization realized the damage from this decision too late.

Situation 3 (The Building an Empire Barrier): This situation has some of the characteristics of Situation 2. In some countries, your salary, power, and authority are based upon the size of the empire you control and the type of decisions you are authorized to make. In such a case, hiring three below-average workers to do the same work as two average workers is better for empire building. Additionally, even though finding adequate human resources may be difficult, sometimes companies simply do not put forth a good search effort; friends and family members may be hired first, regardless of their qualifications. In this situation, the self-serving problem was solved by making a decision that elongated the schedule so that the empire that is built will last as long as possible.

Situation 4 (The Overtime Barrier): Decisions to use overtime can be troublesome. Overtime is usually needed when pressure is placed upon the team to maintain a schedule. However, in some cultures, overtime is used as a reward system to give workers the opportunity to earn additional income. This can occur even if the overtime is not necessary.

Some countries put restrictions on overtime and may require that the government must authorize the overtime especially if it is paid overtime. This occurs when the country is fearful that the overtime, if prolonged, may create a new class of citizens. There is also the danger that mistakes may be made intentionally on a great many projects to justify the use of extended overtime.

Situation 5 (The Career Path Barrier): A government agency discovered that as they began outsourcing more work to national and nonnational contractors, their ability to evaluate project management performance was becoming difficult because each contractor would report status differently. Some contractors appeared to be performing at a higher level of project management than others by using ***PMBOK® Guide*** processes, but the government was unable to compare contractors' performance for awarding future contracts. The government then encouraged all contractors to use project management according to the ***PMBOK®***

Guide and highly recommended that the individuals managing their contracts become certified as PMP®s.

While the government recognized the value in promoting project management professionalism in the contractors' organizations, the government could not see the benefits of professionalism in project management in the public sector and the impact on problem-solving and decision-making. Public sector project managers were treated more so as project monitors than project managers. The Office of Personnel Management in the government was unable to write job descriptions for government project managers because their duties, authority, responsibility, and decision-making capability did not fit in the standard "mold" used for other job descriptions. As such, an assignment as a government project manager was viewed as a nonpromotable position that could impact your career.

Situation 6 (The Certification Barrier): An emerging market nation recognized the benefits of project management implementation and encouraged both government contractors as well as government agencies to support training programs that lead to individuals becoming PMP®s. Training programs were put in place by private firms as well as universities. In most cases, the company or government agency paid for the training of the workers.

Once the individuals became PMP®s, they began asking their companies for a salary increase. The company argued that the cost that the company invested in their education should be viewed as a near-term salary increase and that other financial benefits would occur in the future. The workers were unhappy with the response and had expected immediate financial benefits when becoming certified. As such, the employees found employment in other companies and other countries that would provide them with salary increases. The companies stopped paying for project management training, no longer supported certification efforts, and in many cases discontinued the encouragement to follow the ***PMBOK® Guide***.

Situation 7 (The Educational Barrier): A company that recognized the need for project management implementation sent their workers to training programs. The employees return to work with the expectation that they would be able to implement the tools and techniques they learned in the training. When the employees discovered that their company was reluctant to implement many of the tools and techniques they had learned, the employees sought employment elsewhere.

13.2 Legal Barriers

Some countries establish laws that provide limitations on how much, if any, of the financial resources that the country possesses can leave the country. This creates barriers as shown in Exhibit 13.3. The country can put limitations on procurement

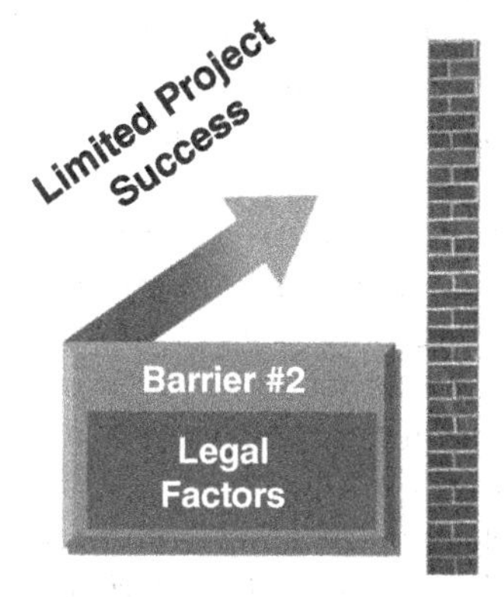

Exhibit 13.3 Legal Barriers

activities that leave the country. There can also be laws on paid overtime for workers.

Companies that wish to do business within these countries must abide by these laws even if the laws seem improper. An example might be the worker's right to hold a job even if the worker's performance is substandard. Some laws may even foster corruption possibilities by making it clear that bribes and "gifts" may be appropriate under certain circumstances during competitive bidding activities.

Situation 8 (The Procurement Barrier): The government in an emerging market nation wanted to limit the procurement of goods and services from outside the country. During competitive bidding, companies were asked to prepare for the government's approval a list of qualified vendors from within the country. Project managers were pressured to use contractors from within the country even if external contractors provided higher-quality goods and services. To make matters worse, additional pressure was imposed to select contractors in cities that had the greatest unemployment rates regardless of the capabilities of the vendors.

Situation 9 (The Unemployment Barrier): A project manager was pressured into awarding a procurement contract to a vendor in a city that had significant unemployment. As the project progressed, the project manager realized she could accelerate the schedule by using overtime. Unfortunately, government permission was required for authorization of overtime pay. The project manager soon discovered that not only would the government not authorize the overtime but was reluctant to allow the project to finish early and with fewer resources which could increase the unemployment level and poverty in the community.

Situation 10 (The Rigid Policies Barrier): A company was awarded a contract for a government agency in a country that utilized rigid official public procurement processes, rules, and laws. The legal environment created a great deal of inflexibility and many of the traditional processes in the ***PMBOK® Guide***, such as with change management activities, were not in line with the government's requirements. To adhere to the inflexibility, the project's budget was increased, and the schedule had to be elongated.

Situation 11 (The Restrictions Barrier): Some countries have government policies that may restrict who a contractor will be allowed to work with, both in the government agency's country and possibly in other countries due to political concerns or competitive factors. Some policies can also dictate who the prime contractor must hire as subcontractors even if the subcontractor's interest in and importance of the project is different from that of the prime contractor.

Even though a customer or government agency follows the processes within the ***PMBOK® Guide***, there can still exist laws or policies related to the criteria for decision-making, the time to make the decision and the parties involved in making the decision. Every government agency can have their own interpretation for the acceptance/rejection criteria for project deliverables, the evaluation of quality, and decisions related to required permits. Some of the differences that can exist between countries include:

- The definition of the level of customer requirements, technical specifications, and quality required since the differences are observed when the client and the contractor are from different countries.
- Timing and requirements for the process of obtaining permits and construction licenses, in accordance with the applicable regulations and laws of the country.
- Interpretation or agreement of requirements or conditions of acceptance of the project and its deliverables.
- Involvement of areas of the organization when negotiating with the client, in terms of scope changes in the project.
- Definition of norms, laws, international treaties, and rules that must be complied with.
- Process of management, administration, and negotiation of leases of machinery and equipment with third parties, when dealing with other countries.
- Be at the forefront in terms of performance and productivity of machinery, equipment, and software existing in the market in other countries and manage to bring them to your project.
- Definition and adequate management of the change control system, through authorized change requests, analysis of impact on risks, time, and project resources.
- Generation of the PMO's value in the organization, which monitors the governance of projects and establishes a common framework of methodologies, processes, policies, and information systems.

13.3 Project Sponsorship Barriers

We assign sponsors or governance committees to projects to provide project teams with a line of sight to senior management for strategic information, assistance for decisions that cannot be made entirely by the project team, coordination of large stakeholder groups, and resolution of problems that can be resolved more effectively by the sponsors. It is not uncommon for individuals to be asked to serve as sponsors without understanding project management, or their role and responsibility as a sponsor, and the fact that project governance is not the same as functional governance. Project problem-solving and decision-making can be different that how it is done in a functional group. There is also the risk that the individual may abuse the position, possibly with hidden agendas. In any event, barriers as seen in Exhibit 13.4 can be created that prevent the implementation of effective project management.

Situation 12 (The Centralization of Authority Barrier):[1] Many countries maintain a culture in which very few people have the authority to make decisions. Decision making serves as a source of vast power in both privately held companies and governmental organizations. Project management advocates decentralization of authority and decision-making. In such countries, the executive level will never surrender its authority and decision-making power to project managers. Project managers may then function as puppets and unable to effectively manage the projects.

Situation 13 (The Lack of Executive Sponsorship Barrier): Project sponsorship may exist somewhere in the company but often not at the executive level,

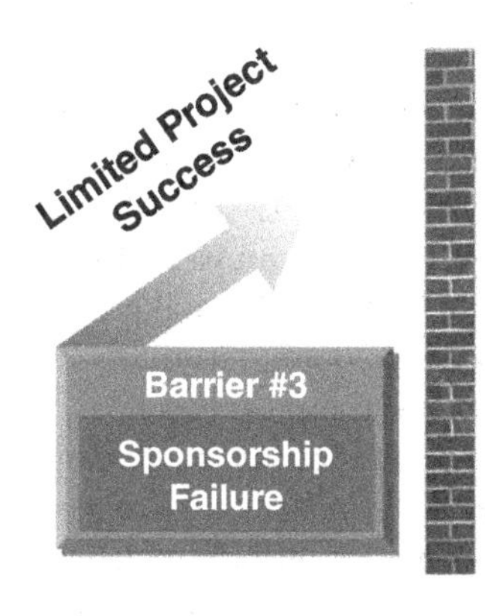

Exhibit 13.4 Sponsorship Barriers

1 Situations 12–16 and 19–22 have been adapted from Kerzner and Zeitoun (2008).

for two reasons. First, executives may recognize that they have no knowledge useful to the project. Therefore, they could make blunders that would become visible to the people that put them in power. Second, and possibly most important, acting as a sponsor on a project that fails could end the executive's career, politically. Therefore, sponsorship, if it exists at all, is often at a low level in the organizational hierarchy, a level at which workers are expendable. The result is that sponsors cannot or will not make decisions or help project managers in times of trouble. We then have invisible sponsorship.

Situation 14 (The Organizational Hierarchy Barrier): In traditional project management practices, we tend to believe that problems are resolved at the project sponsor level. But in nations in which organizational hierarchy is sacred, following the chain of command can elongate the project management process to a point at which schedules become irrelevant. Also, the infrastructure to support project management may exist only to filter bad news from the executive levels and to justify the existence of functional managers. Some decisions and news may go as high as government ministers. Simply stated, project managers may not know where and when a decision will be made and cannot be sure where project information will end up. There may exist excessive bureaucracy that is not visible at the project management level.

Situation 15 (Insecurity at Executive Levels Barrier): Executives may feel insecure about performing as a sponsor because their positions are the result of political appointments. Additionally, project managers may be viewed as the stars of the future and, as such, are a threat to executives. Project management implementation could force the loss of an executive's status, and status is often accompanied by fringe benefits and other privileges. Before executives consider throwing their support behind a new approach, such as project management implementation or a project, they worry about its effect on their power, authority, and chance for advancement.

Situation 16 (The Social Obligations Barrier): In emerging market nations, social obligations due to religious beliefs and politics are a way for executives to maintain alliances with those people who put them in power. As such, project managers may not be allowed to interact socially with certain groups that may possess critical information needed for some decisions. Unlike traditional project management practices where PMs may have the right to communication with everyone, it may not be possible without going through the project sponsor.

Situation 17 (The Lack of Education Barrier): Not all sponsors understand project management or have a desire to attend courses on project management. Being unsure about their role as a sponsor, they focus mainly on how to deliver the results faster and at a lower cost regardless of the quality, risks, or the best way to achieve the results. Sometimes, sponsors will make promises to customers for rework or additional testing at no cost to the customer. This creates havoc for project teams that need effective decisions to be made.

Situation 18 (The Project Charter Barrier): Without at least a cursory knowledge of project management, sponsors are reluctant to prepare and sign a project charter for fear that the costs and schedule are not well estimated. This then forces some team members and even stakeholders to begin to perform some of the tasks without prior authorization.

13.4 Cost of Implementation Barriers

Most organizations today understand the benefits that can be forthcoming from effective project management implementation but are unsure about the costs associated with obtaining the benefits. This then creates issues when implementation problems must be resolved and the types of decisions to be made. As such, there may be apprehension or even fear in a commitment, and the creation of a barrier as shown in Exhibit 13.5 will occur.

The fear is often that readers of the 756-page ***PMBOK® Guide*** believe that all the processes, inputs, outputs, tools, and techniques must be implemented for the benefits to be recognized. This is not the case. There are extensions to the ***PMBOK® Guide*** where it can be downsized for specific needs. This could be the reason why the 18-page ***SCRUM Guide*** has become popular.

Situation 19 (The Cost of Implementation Barrier): The costs associated with the implementation of project management include purchasing hardware and software, creating a project management methodology, and developing project performance reporting techniques. Costs such as these require a significant financial expenditure that the company may not be able to afford. They also require tying up significant resources in implementation for an extended time. With limited resources and better human resources – which are required for

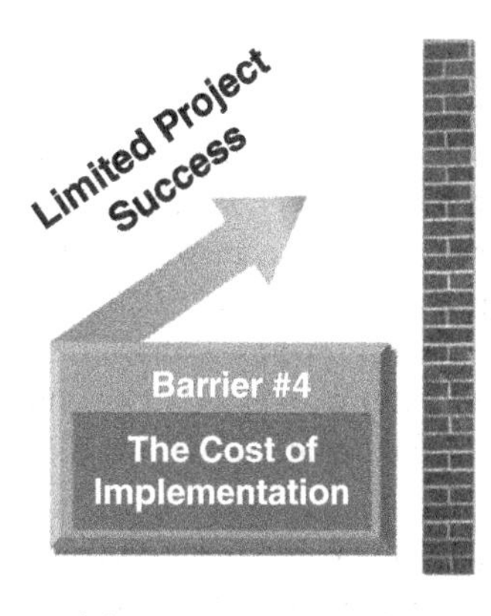

- Not everyone is willing to commit the funds for the tools, techniques and processes needed for project management implementation even though the benefits are known.

Exhibit 13.5 Implementation Barriers

implementation – removed from ongoing work, companies may avoid project management even when they recognize its benefits.

Situation 20 (The Risk of Failure Barrier): Even if a company is willing to invest the time and money for project management implementation, there is a significant risk that implementation will fail. And even if implementation is successful but a project begins to fail, for any number of reasons, blame may be placed on faulty implementation. Executives who must then explain the time and money expended for no real results may find that their position in the hierarchy is now insecure. Therefore, some executives refuse to either accept or visibly support project management.

Situation 21 (The Costs of Training Barrier): Project management implementation is difficult without training for the workers. This includes project managers, team members, and sponsors. The need for training can create additional problems. First, how much money must be allocated for training? Second, who will provide the training and what are the credentials of the trainers? Third, can people be released from current project work to attend training classes? It is time-consuming and expensive to train people in project management. Adding together the cost of implementation and the cost of training may frighten executives away from accepting project management.

Situation 22 (The Need for Sophistication Barrier): Project management requires sophistication, not only in the technology but also in the ability of people to work together. In emerging market countries, employees may not have been trained for teamwork and may not be rewarded for their teamwork contributions. Communication skills may be weak, including report writing. People may see teamwork as a medium through which others are able to recognize their lack of competencies and mistakes.

13.5 Culture Barriers

Sustainable project management success requires a cooperative culture where team members work together and make decisions in the best interest of the firm. This is most often accomplished without considering authority, power, or pay grade. Cooperative cultures often determine the type of organizational structure to be used for projects. As an example, the matrix structure seems to work best with cooperative cultures. Yet even within cooperative cultures, barriers as shown in Exhibit 13.6 can be erected if the employees feel threatened or have a hidden agenda.

Situation 23 (The Planning Barrier): If an organization lacks standards or commitment for project management, then the planning process may struggle with poor estimates for work effort, duration, and cost. The cultural barrier can

Exhibit 13.6 Cultural Barriers

occur if the organization fails to support the use of a methodology. The result can be ambiguous scope and requirements. Poor planning usually translates into plans that change too often, unrealistic milestones, and a lack of faith at all levels of management that project management can succeed. If there are risks that the plans may fail, workers may create barriers and find excuses as to why they cannot participate in planning activities.

Situation 24 (The Paperwork Nightmare Barrier): When looking at the number of activities in the ***PMBOK® Guide***, there is apprehension that an excessive amount of unproductive time may be spent completing all the paperwork required. A barrier can then be created with a multitude of reasons why certain reports and paperwork are not necessary. The barrier can be significant if any organization believes in the philosophy that "what is not on paper has not been said."

Situation 25 (The Project Completion Barrier): As projects begin to wind down, workers begin to worry about their next assignment. If they are unsure about their next assignment, they may elongate the project closure process. They may also leave their current project before the work is finished to guarantee employment elsewhere. This may create a hardship for the remaining team members.

Sometimes, there exists a lack of dedication to project closure. Employees are often afraid to be attached to the project at closure when lessons learned and best practices are captured. Lessons learned and best practices can be based upon what was done well and what was done poorly. Employees may not want anything in writing that indicates that best practices were discovered through their mistakes.

Situation 26 (The Management Reserve Barrier): Cooperative cultures tend to find ways to protect themselves and their colleagues from some risks, rework, poor estimating, and other such situations. The management reserve is one such way. However, a barrier can be created when the customer believes that the management reserve is created solely for the benefit of the contractor rather than the customer as well.

Situation 27 (The PMP® Barrier): Not all organizations encourage their people to become certified as project management professionals. One of the benefits of certification is that it becomes easier to understand each person's role and responsibility, thus providing a good foundation for a cooperative culture. In noncooperative cultures, the role of the project manager becomes that of a firefighter.

13.6 Project Management Office (PMO) Barrier

Even though there are several types of PMOs, the existence of a PMO usually implies that there exists an organization dedicated to continuous improvements in project management. The PMO can also participate in many of the decisions that must be made as well as establishing documentation requirements for problem-solving and decision-making. Unfortunately, executives may view the PMO as a threat if the executives are happy with the status quo and their position in the firm and do not want to see any changes take place. This barrier is shown in Exhibit 13.7.

PMOs can be used not only to promote effective project management practices but also to capture and share best practices and lessons learned as well as validate some decisions. They can also assist senior management in the monitoring of enterprise environmental factors that can impact decision-making. This is important in countries that struggle with hyperinflation and must control the country's scarce resources rapidly.

- **This barrier occurs when organizational leadership does not believe that project management implementation is in their best interest. Personal goals become more important than organizational goals.**

Exhibit 13.7 The PMO Barrier

Situation 28 (The "Information is Power" Barrier): When managers believe that information is power, they create barriers that may prevent the formation of organizations that gather and disseminate information. Sometimes, several functional PMOs are created to centralize functional information and control its release. Infighting can occur in functional units as to who controls each PMO if the information is seen as a source of power.

Situation 29 (The PMO Financing Barrier): All PMOs require human and nonhuman resources to be effective. Resources require capital expenditures. Managers that believe information is power will always create barriers that justify not financing a PMO. Barriers can be established by either staffing the PMO with nonqualified individuals or limiting the tools provided to the PMO for monitoring and controlling projects.

13.7 Conclusion

All countries, including emerging market nations, have an abundance of talent that has yet to be fully harvested. Barriers can appear in any organization for a multitude of reasons as identified in the situations presented. Virtual project management teams as well as a recognition of the benefits of project management may be the starting point for the full implementation of project management.

As project management grows, executives will recognize and accept the benefits of project management and see their business base increase. Partnerships and joint ventures may become more prevalent. The barriers that impede successful project management implementation will still exist, but we will begin to excel in how to live and work within the barriers and constraints imposed.

Corporate executives worldwide are beginning to see more of the value of project management and have taken steps to expand its use. Executives in some of the rapidly developing nations appear today to be much more aggressive in providing the support needed for breaking through many of the barriers. As more success stories emerge, we will see the various economies strengthening and becoming more connected, and executives will start to implement project management more fully.

Discussion Questions

1. Why do barriers to project management implementation exist?
2. Can the barriers impact effective problem-solving and decision-making?
3. What can the project manager do to overcome barriers that impact effective problem-solving and decision-making?
4. Will the barriers ever completely disappear?

References

Ali, U. and Kidd, C. (2014). Barriers to effective configuration management application in a project context: an empirical investigation. *International Journal of Project Management* 32 (3): 508–518. 11p. DOI: https://doi.org/10.1016/j.ijproman.2013.06.005.

Ambekar, S. and Hudnurkar, M. (2017). Factorial structure for Six Sigma project barriers in Indian manufacturing and service industries. *TQM Journal* 29 (5): 744–759. 16p. DOI: https://doi.org/10.1108/TQM-02-2017-0021.

Arnold, P. and Javernick-Will, A. (2013). Projectwide access: key to effective implementation of construction project management software systems. *Journal of Construction Engineering and Management* 139 (5): 510–518. 9p. DOI: https://doi.org/10.1061/(ASCE)CO.1943-7862.0000596.

Blixt, C. and Kirytopoulos, K. (2017). Challenges and competencies for project management in the Australian public service. *International Journal of Public Sector Management* 30 (3): 286–300. 15p. DOI: https://doi.org/10.1108/IJPSM-08-2016-0132.

Carvalho, M.M.d. (2014). An investigation of the role of communication in IT projects. *International Journal of Operations & Production Management* 34 (1): 36–64. 29p. DOI: https://doi.org/10.1108/IJOPM-11-2011-0439.

Hwang, B.-G. and Tan, J.S. (2012). Green building project management: Obstacles and solutions for sustainable development. *Sustainable Development* 20 (5): 335–349. 15p. 3 Charts. DOI: https://doi.org/10.1002/sd.492.

Hwang, B.-G., Zhao, X., and Toh, L.P. (2014). Risk management in small construction projects in Singapore: status, barriers and impact. *International Journal of Project Management* 32 (1): 116–124. 9p. DOI: https://doi.org/10.1016/j.ijproman.2013.01.007.

Hwang, B.-G., Zhu, L., and Tan, J.S.H. (2017). Green business park project management: barriers and solutions for sustainable development. *Journal of Cleaner Production* 153: 209–219. 11p. DOI: https://doi.org/10.1016/j.jclepro.2017.03.210.

Johansen, J. and Gillard, S. (2005). Information resources project management communication: personal and environmental barriers. *Journal of Information Science* 31 (2): 91–98. 8p. DOI: https://doi.org/10.1177/0165551505050786.

Kerzner, H. and Zeitoun, A. (2008). Barriers to implementing project management in emerging market nations. International Institute for Learning (IIL) white paper.

Khan, A.A. and Keung, J. (2016). Systematic review of success factors and barriers for software process improvement in global software development. *IET Software* 10 (5): 125–135. 11p. DOI: https://doi.org/10.1049/iet-sen.2015.0038.

Khan, S.U., Niazi, M., and Ahmad, R. (2011). Barriers in the selection of offshore software development outsourcing vendors: an exploratory study using a systematic literature review. *Information and Software Technology* 53 (7): 693–706. 14p. DOI: https://doi.org/10.1016/j.infsof.2010.08.003.

Kutsch, E. and Hall, M. (2010). Deliberate ignorance in project risk management. *International Journal of Project Management* 28 (3): 245–255. 11p. DOI: https://doi.org/10.1016/j.ijproman.2009.05.003.

Kutsch, E. and Hall, M. (2009). The rational choice of not applying project risk management in information technology projects. *Project Management Journal* 40 (3): 72–81. 10p. 4 Charts. DOI: https://doi.org/10.1002/pmj.20112.

Loushine, T.W., Hoonakker, P.L.T., Carayon, P., and Smith, M.J. (2006). Quality and safety management in construction. *Total Quality Management & Business Excellence* 17 (9): 1171–1212. 42p. 1 Diagram, 4 Charts. DOI: https://doi.org/10.1080/14783360600750469.

Magnier-Watanabe, R. and Benton, C. (2013). Knowledge needs, barriers, and enablers for Japanese engineers. *Knowledge and Process Management* 20 (2): 90–101. 12p. 5 Charts, 1 Graph, 6 Maps. DOI: https://doi.org/10.1002/kpm.1408.

Moore, D.R. and Dainty, A.R.J. (2001). Intra-team boundaries as inhibitors of performance improvement in UK design and build projects: a call for change. *Construction Management and Economics* 19 (6): 559–562. 4p. DOI: https://doi.org/10.1080/01446190110055508.

Niazi, M., Babar, M.A., and Verner, J.M. (2010). Software Process Improvement barriers: a cross-cultural comparison. *Information and Software Technology* 52 (11): 1204–1216. 13p. DOI: https://doi.org/10.1016/j.infsof.2010.06.005.

Paterson, C.J. and Andrews, R.N.L. (1995). Procedural and substantive fairness in risk decisions: comparative risk assessment procedures. *Policy Studies Journal* 23 (1): 85–95. 11p. DOI: https://doi.org/10.1111/j.1541-0072.1995.tb00508.x.

Polak, J. and Wójcik, P. (2015). Knowledge management in IT outsourcing/offshoring projects. *PM World Journal* 4 (8): 1–10. 10p.

Sakellariou, E., Karantinou, K., and Poulis, K. (2013/2014). Managing the global front end of NPD: lessons learned from the FMCG industry. *Journal of General Management* 39 (2): 61–81. 21p. DOI: https://doi.org/10.1177/030630701303900204.

Santos, V.R., Soares, A.L., and Carvalho, J.Á. (2012). Information management barriers in complex research and development projects: an exploratory study on the perceptions of project managers. *Knowledge and Process Management* 19 (2): 69–78. 10p. DOI: https://doi.org/10.1002/kpm.1383.

Senesi, C., Javernick-Will, A., and Molenaar, K.R. (2015). Benefits and barriers to applying probabilistic risk analysis on engineering and construction projects. *Engineering Management Journal* 27 (2): 49–57. 9p. 5 Charts. DOI: https://doi.org/10.1080/10429247.2015.1035965.

Sommer, A.F., Dukovska-Popovska, I., and Steger-Jensen, K. (2014). Barriers towards integrated product development – challenges from a holistic project management perspective. *International Journal of Project Management* 32 (6): 970–982. 13p. DOI: https://doi.org/10.1016/j.ijproman.2013.10.013.

Terlizzi, M.A., Meirelles, F.d.S., and de Moraes, H.R.O.C. (2016). Barriers to the use of an IT Project Management Methodology in a large financial institution. *International Journal of Project Management* 34 (3): 467–479. 13p. DOI: https://doi.org/10.1016/j.ijproman.2015.12.005.

Tang, W., Qiang, M., Duffield, C.F. et al. (2007). Risk management in the Chinese construction industry. *Journal of Construction Engineering and Management* 133 (12): 944–956. 13p. 1 Diagram, 8 Charts. DOI: https://doi.org/10.1061/(ASCE)0733-9364(2007)133:12(944).

Appendix: Using the *PMBOK® Guide*

Decision-Making and the *PMBOK® Guide*

Logical decision-making is an important part of all project management, and this is a necessity for all of the ***PMBOK® Guide*** Domain Areas and Areas of Knowledge. The ***PMBOK® Guide*** provides a reasonable roadmap for what must be done on typical projects.

The ***PMBOK® Guide*** is extremely useful for helping us make informed decisions, especially with the identification of "inputs," "tools and techniques," and "outputs." However, in situations involving time pressure, higher stakes, or increased ambiguities, project managers often use intuitive decision-making practices rather than structured approaches and generally arrive at a satisfactory course of action, possibly without weighing alternatives. Because of time constraints, decisions may have to be made with just partial information being available.

Yet even under these circumstances, people familiar with the ***PMBOK® Guide*** can be expected to make better decisions than those unfamiliar with it. The ***PMBOK® Guide*** does not and most likely cannot provide guidance on how to approach every possible problem that can exist and how to make every possible decision. But having some information on the ***PMBOK® Guide*** is certainly better than having no information at all. While some structured approaches exist for decision-making, they are not applicable to every type of project problem.

Problem-Solving and the *PMBOK® Guide*

The ***PMBOK® Guide*** is an excellent document for learning project management but as the name implies, it is still a "Guide." The ***PMBOK® Guide*** cannot possibly

Project Based Problem Solving and Decision Making: A Guide for Project Managers,
First Edition. Harold Kerzner.

Companion Website: www.wiley.com/go/kerzner/projectbasedproblemsolving

list all of the possible problems that can occur on a project and all of the alternatives and decisions that must be made.

Every project has its unique characteristics and problems. Some companies require that project managers maintain diaries on projects. New or inexperienced project managers may be required to read these diaries if they are managing similar projects. These diaries usually discuss what problems occurred, such as risks, and what the project manager did to resolve the problems and mitigate the risks.

Not all problems carry the same level of importance. Some problems must be resolved quickly whereas others may have the luxury of having the solution delayed a bit. Not all of the problems necessarily lead to project failure, but they can create severe headaches during the execution of the project. Some of these problems may not be able to be resolved and the project team must simply adopt a "let's-live-with-it" attitude.

In the next, several sections are listed some of the common problems that can occur in the knowledge areas of the ***PMBOK® Guide.*** Once again, these are just partial lists of problems that can occur and impact problem-solving and decision-making.

PMBOK® Guide: Integration Management

Potential problems in Integration Management that may require decision-making include:

- Unclear business case or lack of a business case
- Ill-defined statement of work
- No identification of the enterprise environmental factors that can impact project decision-making
- The customer and the stakeholders want status information that is not readily available through the normal organizational process assets
- Deciding whether scope changes should be done incrementally throughout the project or performed later as an enhancement project
- Verification and validation failures during project or life cycle phase closure activities
- Stakeholders are using different organizational process assets which cannot interface with each other.

PMBOK® Guide: Scope Management

Potential problems in Scope Management that may require decision-making include:

- Not being able to identify all of the stakeholders and their needs
- Having stakeholders with different or competing requirements

- Not being able to get stakeholder agreement on the project's requirements or a course of action
- Unable to decide which tools to use for collecting requirements
- Finding errors in the traceability requirements matrix
- Not enough detail in the work breakdown structure
- Too much detail in the work breakdown structure such that functional managers and people doing the activities have difficulty in reporting status
- Constantly changing milestones
- Having team members that may be inexperienced with project planning
- No systemization of the planning process
- Having technical objectives that appear to be more important than business objectives.

PMBOK® Guide: Time Management

Potential problems in Time Management that may require decision-making include:

- Customer requested changes in the end date of the project (compression or elongation)
- Using the wrong estimating techniques for duration estimates
- Agreeing to an unrealistic end date just to win the contract
- Determining the best approach for schedule compression
- Difficulty in determining the activity list from a WBS that is too detailed
- Unable to accurately estimate activity resources
- Project estimates are best guesses and not based upon history or standards
- Difficulty in determining which tools and techniques are best for estimating durations
- Having too many leads and lags associated with the schedule
- Unable to come up with a reasonable contingency plan
- Having poorly written requirements that mandate the use of rolling wave scheduling
- Having a team that has difficulty in performing what-if scenarios
- Not having all of the necessary inputs to develop the schedule.

PMBOK® Guide: Cost Management

Potential problems in Cost Management that may require decision-making include:

- Too many cost-estimating techniques and each one produces different results
- Having an estimating group that estimate all work and the team does not believe that they can do the work according to the estimates provided

- Constantly changing resources that can affect the cost estimates
- Estimating based upon incomplete requirements
- Budgets are being exceeded and out of control
- The sales force commits to making "no-cost" scope changes for the customer, but the additional costs are actually significant
- Not enough time was allocated for proper estimating to take place
- Budgeting did not account for a management reserve
- The management reserve is being consumed rapidly because of escalation factors.

PMBOK® Guide: Quality Management

Potential problems in Quality Management that may require decision-making include:

- Being unfamiliar with the tools and techniques that are used for quality planning
- Being unfamiliar with the tools and techniques that are used for quality control
- Using the wrong tools and techniques
- Having no templates or checklists for quality planning
- Having a customer and stakeholders that seem to have more knowledge about quality than does the project team
- Having no quality metrics
- Not performing quality audits
- Promising the customer a higher level of quality than you can deliver
- Failing to understand the importance of quality during validation and verification activities.

PMBOK® Guide: Human Resource Management

Potential problems in Human Resource Management that may require decision-making include:

- Having a lack of resources
- Being assigned resources that lack experience
- Not fully knowing the staffing requirements
- Having constantly changing resources
- Having people shuffled in and out of the project with little regard for the schedule
- Lack of attention provided to the human and organizational aspects of the project

- Working with a burned-out team
- Asking people to work excessive overtime
- Having no staffing plan
- Having no recognition and rewards system on the project
- People are part of a virtual team but do not understand how virtual teams should work
- Unsure how to motivate team members
- Too many conflicts that go unresolved.

PMBOK® Guide: Communications Management

Potential problems in Communications Management that may require decision-making include:

- End-user stakeholders are not being involved in the project
- Minimal or no stakeholder backing; lack of ownership
- Weak project and stakeholder communications
- Not knowing which stakeholders are critical and have the power to cancel the project
- Not understanding the different types of performance reports
- Not establishing a communications matrix
- Having to repeat instructions because of communications barriers
- Not understanding how to conduct an effective team meeting
- Lack of feedback which results in poor understanding of instructions
- Selecting the wrong media for communicating something critical
- Receiving improper instructions in the use of proprietary information
- Not understanding the impact of personality and perception screens/barriers
- Not knowing which team members have the authority to make decisions for their line or unit managers.

PMBOK® Guide: Risk Management

Potential problems in Risk Management that may require decision-making include:

- Having no risk management plan even though the customer expected a plan to be developed
- Not being familiar with the tools and techniques available for identifying risks
- Having a project team that has a poor understanding of risk management and its importance

- Having no strategies for unfavorable risks
- Having no strategies for favorable risks that provide added opportunities
- Recognizing the risks but not knowing how to respond and control the risks
- Not knowing about the risks of previous projects, especially those risks that could have an impact on your current project
- Not identifying risk triggers
- Not having a risk register
- Not allocating sufficient funds for risk management activities
- Having clients and stakeholders that have a poor understanding of risk management.

PMBOK® Guide: Procurement Management

Potential problems in Procurement Management that may require decision-making include:

- Deciding between the make or buy option
- Deciding upon the best contract type
- Failing to understand risk-related contract decisions
- Deciding how scope changes may impact the terms and conditions of the contract
- Not understanding how the client's payment plan impacts your cash flow
- Not understanding how the project's procurement requirements impact cash flow
- Discovering during bidders' conferences that your proposal did not include critical information
- Having to decide between single-source and multiple-source subcontracting
- Failing to adhere to customer's request for life cycle costing efforts
- Failing to consider ramification of noncompete and nondisclosure clauses
- Failing to consider waivers
- Failing to account for resolution of disputes in the contract
- Failing to account for a change management process.

PMBOK® Guide: Stakeholder Management

Potential problems in Stakeholder Management that may require decision-making include:

- Failing to identify all the stakeholders
- Not knowing which stakeholders are important

- Not knowing which stakeholders have authority to participate in decision-making
- Failing to create a power-influence grid
- Not knowing which stakeholders simply prefer to be informed
- Not knowing how frequently to communicate with stakeholders
- Failing to identify if any stakeholders might have hidden agendas
- Failing to include stakeholder engagement as part of the project's communications plan.

Further Reading

Buchannan, L. and O'Connell, A. (2006). A brief history of decision making. *Harvard Business Review:* 32–41.

Drucker, P. (2008). Managing oneself. *Harvard Business School Books:* 1–72.

Einhorn, C.S. (2021). Myths about decision-making. *Harvard Business Review:* 1–7.

Einhorn, C.S. (2022). What are your decision-making strengths and blind spots? *Harvard Business Review.*

Greg, Z.F. (2009). *How to Solve Any Problem.* Lexington, KY: Cexino.

Hammond, J.S., Keeney, R.L. and Raiffa, H. (2023). The hidden traps in decision making. *Harvard Business Review:* 62–71.

Higgins, J.M. (2006). *101 Problem Solving Techniques.* Winter Park, FL: New Management Publishing.

Morgan, D.J. (1998). *The Thinker's Toolkit.* New York: Times Books.

Moore, M.G. (2022). How to make great decisions quickly. *Harvard Business Review.*

Proctor, T. (1999). *Creative Problem Solving for Manager,* 3e. United Kingdom: Routledge.

Snowden, J.D. and Boone, M.E. (2007). A leader's framework for decision making. *Harvard Business Review:* 149–154.

Vaughn, R.H. (2007). *Decision Making and Problem Solving in Management.* Brunswick, OH: Crown Publishers.

Watanabe, K. (2009). *Problem Solving 101.* New York, NY: Penguin Group.

William, J.A. (1999). *The Thinking Manager's Toolbox.* New York: Oxford Press.

Project Based Problem Solving and Decision Making: A Guide for Project Managers,
First Edition. Harold Kerzner.

Companion Website: www.wiley.com/go/kerzner/projectbasedproblemsolving

Index

Project Based Problem Solving and Decision Making: A Guide for Project Managers,
First Edition. Harold Kerzner.

Companion Website: www.wiley.com/go/kerzner/projectbasedproblemsolving

D

L

M

N

O

P

T

U

V

W